100 clever ways to help you understand and remember the most important theories

RICH COCHRANE

An Hachette UK Company
www.hachette.co.uk

First published in Great Britain in 2018 by Cassell, a division of
Octopus Publishing Group Ltd.
Carmelite House
50 Victoria Embankment
London EC4Y 0DZ
www.octopusbooks.co.uk
www.octopusbooksusa.com

Distributed in the US by
Hachette Book Group
1290 Avenue of the Americas
4th and 5th Floors
New York, NY 10104

Distributed in Canada by
Canadian Manda Group
664 Annette St.
Toronto, Ontario, Canada M6S 2C8

ISBN 978-1-78840-012-1

A CIP catalog record for this book is available from the British Library.

Printed and bound in China

10 9 8 7 6 5 4 3 2 1

Publishing Director: Trevor Davies
Senior Editor: Pauline Bache
Junior Designer: Jack Storey
Design and layout: Simon Buchanan, Design 23
Illustrators: Design 23
Copyeditor: Mairi Sutherland
Production Controller: Sarah Kulasek-Boyd

Contents

Introduction

Why Math Hacks?

There is an ancient story that goes like this: King Ptolemy I of Egypt had engaged the famous geometer Euclid as his private tutor but quickly became frustrated by the difficulty of the subject and how long it was taking to make progress. Surely, he put it to his teacher, there is a quicker way? A shortcut? A hack, perhaps? "There is no royal road to geometry," Euclid replied firmly.

It probably didn't happen quite like that, but the conversation has certainly been had countless times since. Euclid's answer is broadly right, and it applies not only to mathematics. Many a music student has complained about seemingly endless hours practicing scales, and budding athletes have similar grievances. Learning something hard *is* hard—if it weren't, everyone would do it.

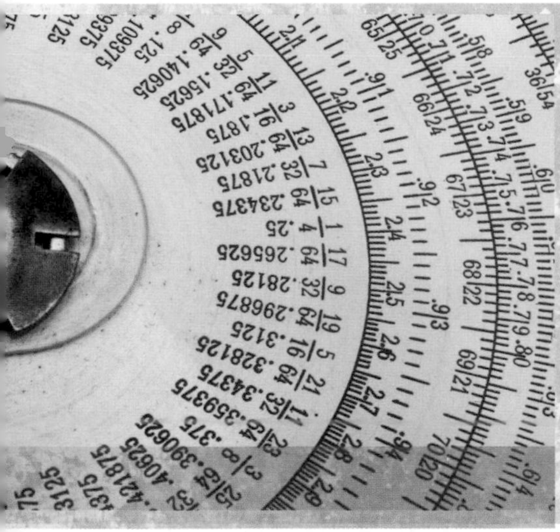

There may not be a royal shortcut, but if you are planning a road trip into mathematics there are better and worse ways to prepare. One thing you should probably have is a map that points out the features you might want to visit and how to get from one to another.

That is primarily what this book is; a tourist's gazetteer of mathematics. The subject's size and scope can be daunting to a visitor, who is liable to get lost, especially if they don't have even a smattering of the local language. Like all good guidebooks, this doubles as a basic phrasebook, and it presents an opinionated, biased, and personal view. If a purely objective picture is possible, which I doubt, you won't find it here.

The map is not the territory, and reading this book will not make you a mathematician. It will, however, give you a sense of what math is and the kinds of things it studies. Almost certainly these are quite different from your school experience, where you were probably made to do the equivalent of memorizing the lengths of rivers and the names of capital cities; trivial, grinding, book work. Real mathematics is more about the journey than where you arrive (nobody ever "arrives" anyway; everyone is a student, a traveler).

When you visit a city, it's nice to know when the cathedral was built and by whom, but only if it's still standing. It's also important to know how the subway works and where the good hotels are. So, although it contains some historical material, this book is primarily a guide to today's field. I have tried to ensure all major strands of contemporary pure mathematics are represented, and to include some of the most important and dramatic results from the last century. This sometimes means covering topics that are intrinsically "advanced" and that require more preparation than this book can reasonably provide. This book cannot really teach you what homological algebra is, for example, but it can tell you it exists, and roughly where it is on the map. These topics are like mountains, and you will need more than a guidebook if you intend to climb them. Here you will discover where they are and get a hint of why you might consider a hike one day.

Parts of the Book

We start with "Tricks of the Trade," which are ideas and techniques that pervade almost all of mathematics. Part 2 is on "Numerous Numbers," the things most folks think mathematics is all about. The idea of number itself has been radically reimagined over the last two centuries. Mathematics is actually about much more than numbers. One plausible claim is that it is "The Science of Structure," which is the focus of Part 3.

Parts 4 and 5 pick up on a different but closely related strand, broadly, mathematics as the study of space and time. In "Continuity" we look at calculus, a family of techniques for studying processes of change and other continuous phenomena that have undergone a vast generalization since their invention by Newton and Leibniz.

In "Math in Space" we see how geometry has also evolved into a rich field populated by strange and exotic objects. I have restrained myself from describing things like the Möbius strip, which are discussed in almost every popular mathematics book. Here we go quite a bit deeper, visiting topology and Riemannian and algebraic geometries.

Finally, in "Math Meets Reality" I try to do some justice to the areas of mathematics that have mostly evolved in relation to practical applications, especially around statistics, algorithms, decision-making, and modeling. I look at these from a mathematical viewpoint, though, not a scientific one.

Features

Each of the 100 sections aims to give you a general, intuitive sense of the subject. It presents the material in different ways in the hope that one of them works for you. Usually the *Helicopter View* provides some context for the idea and perhaps a motivating problem or example. The *Shortcut* tends to give more specific details (I rarely venture to give what a mathematician would call a "definition," but the intention is similar). Sometimes, however, the topic under discussion seemed to demand a different division of duties between these subsections, so the *Hack* at the end gives you two different, brief ways to remember the idea. They might also jog your memory if you need a quick refresher. It is tough to keep everything straight in your head, especially at first, so this sort of thing can be more helpful than you might expect.

Two of the most important features of the book are the index and the cross-references. Mathematics is an intricately interconnected subject and no part is really disjointed from the others. It is completely normal when learning about something new to have to scurry back and forth between different topics. The more you learn, the easier it gets, although of course it never gets *easy*. Where would the fun be in that?

No.1

Euclid (c.323–283 B.C.E.)

Axiom, Theorem, Proof

The mathematician's minimalist style

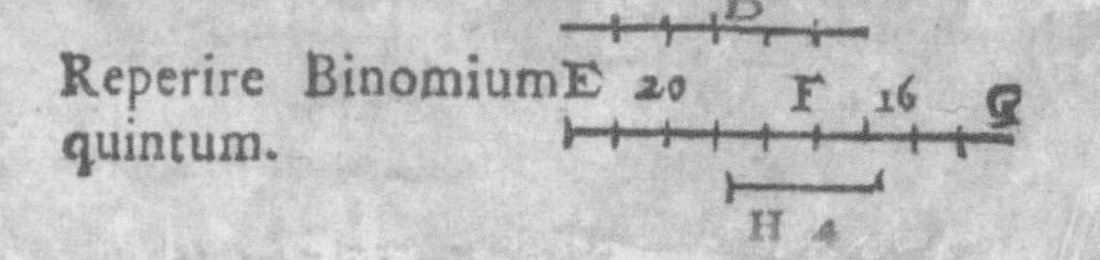

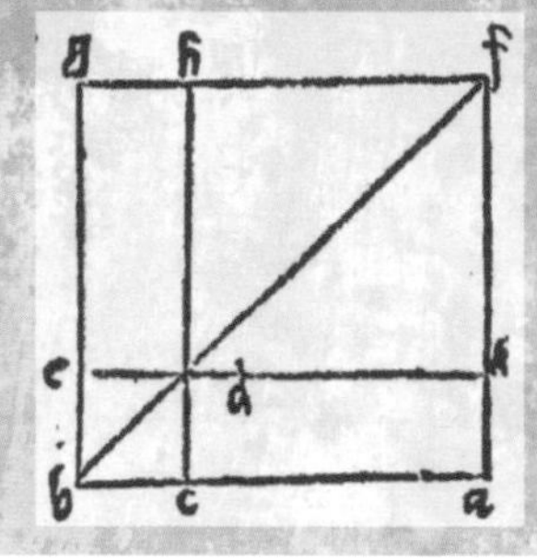

1/ Helicopter view: Euclid wrote his mammoth book *Elements* around 300 B.C.E. It is a collection of mathematical facts, mostly geometrical, which has become one of the most widely read books of all time.

Euclid's book is remarkable for its format as well as its contents. Almost everything in the book belongs to one of three categories. Today these are usually called axioms, theorems, and proofs. They make clear what must be assumed from the beginning, what can be proved from those assumptions, and which methods are used to obtain those results.

Euclid's approach has been copied and adapted by mathematical writers ever since, especially for technical texts. In the 20th century, in particular, a very pared-down version developed that has since become the standard. Some form of the axiom, theorem, proof style is now normal in everything from textbooks to research papers.

Mathematical research often involves proving new theorems from an existing set of axioms; sometimes mathematicians invent whole new sets of axioms, too.

Right: Pages from various editions of Euclid's *Elements*, the most influential book on math ever written.

2/Shortcut:

A mathematical *theory* is the collection of all the facts you can prove from a given set of starting assumptions. *Axioms,* also often called "definitions," are those starting assumptions. They characterize the particular theory you are working in.

If you can argue from those axioms to reach a conclusion that wasn't explicit in them already, that conclusion is called a *theorem* and the argument used to reach that conclusion is the *proof*. Inspecting the proof allows anyone to verify that, if the axioms are true, your theorem must be true, too.

See also //
7 Set Theory, p.18
13 Categories, p.30
14 Natural Numbers, p.32

3/Hack:

Much modern math works by adopting a set of axioms and seeing what theorems can be proved from them, or sometimes inventing new axioms.

Assume the axioms to prove the theorems.

No.2

Induction

Proof by chain reaction

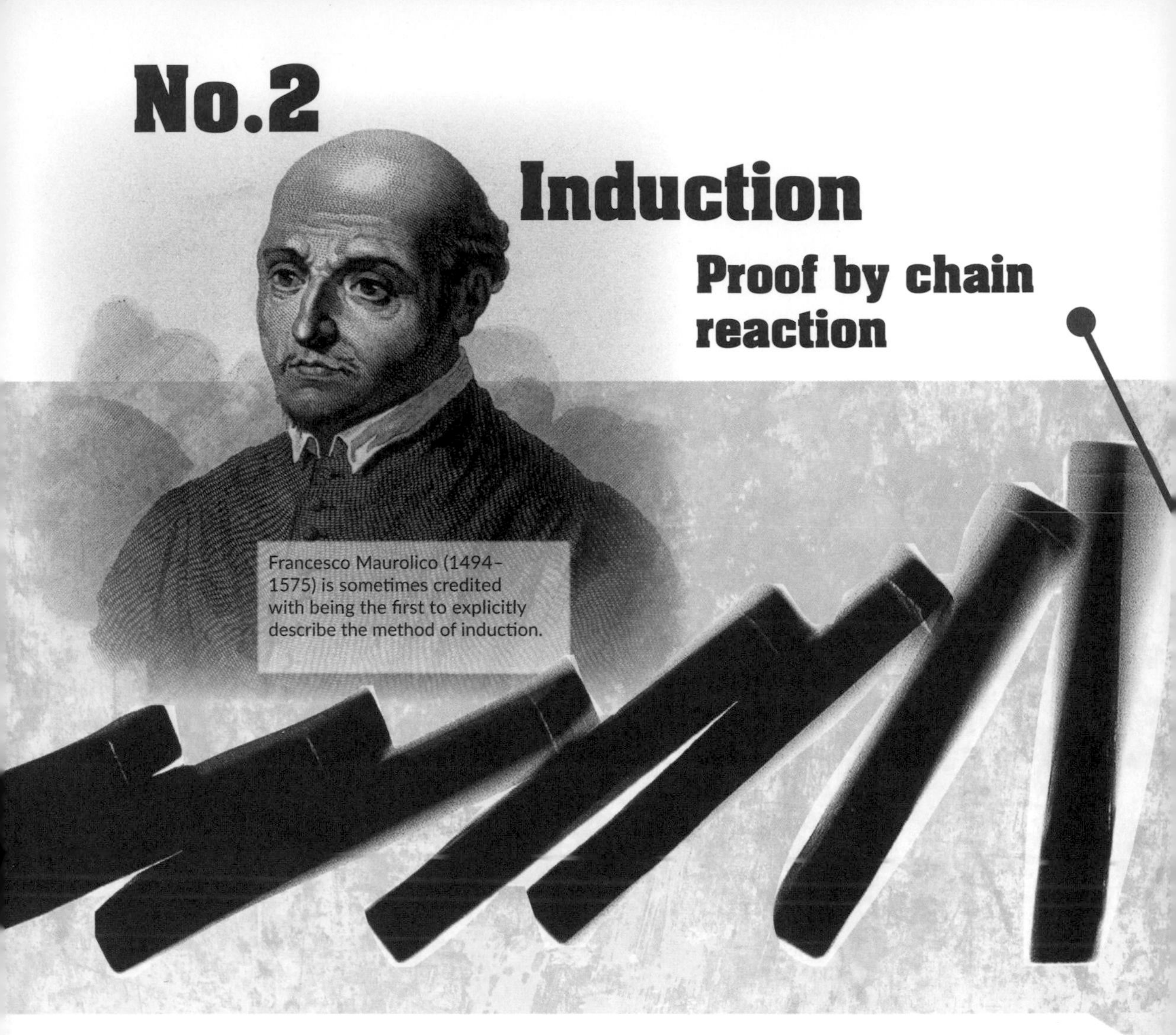

Francesco Maurolico (1494–1575) is sometimes credited with being the first to explicitly describe the method of induction.

1/ Helicopter view: Suppose you want to prove that $n^2 > n$ for every **natural number** n greater than 1. Imagine an infinitely long chain of dominoes waiting to be knocked over. The first is $n = 2$, the next is $n = 3$, and so on. A domino only falls down if we can prove $n^2 > n$ for the value of n it represents.

Our aim is to knock them all down. We *could* try to prove that $2^2 > 2$, then that $3^2 > 3$, and so on, knocking them down one by one, but we'd never get finished. Instead we try something cunning.

First we prove that the first domino falls (prove it for $n = 2$, in our example). Second we prove that if one domino falls, so does the one next to it. If so, then every domino must eventually fall. This is a *proof by induction.*

Right: Bob Speca topples a record-breaking 111,111 dominoes by pushing over a single one; mathematicians can manage an infinite number.

2/Shortcut: The *base case* is a version of what we want to prove that applies only to the smallest number. In this case, it's the claim that $2^2 > 2$. But $2^2 = 4$, and $4 > 2$, so the base case is true. This knocks down the first domino.

The *induction step* says that if the statement is true for any n, it's true for $n + 1$. This says that if domino n falls, so does domino $n + 1$. In this case, a bit of algebra tells us that, indeed, $(n + 1)^2 > n + 1$ whenever $n^2 > n$. Each domino that falls knocks over the next one.

See also //
5 Logic, p.14
14 Natural Numbers, p.32
89 Iteration, p.182

3/Hack: Induction can prove an infinite number of facts in a finite time if they can be arranged in an ordered sequence.

The base case knocks down the first domino; induction says each domino knocks down the one after it.

No.3

Reductio ad Absurdum

If it can't not be, it must be

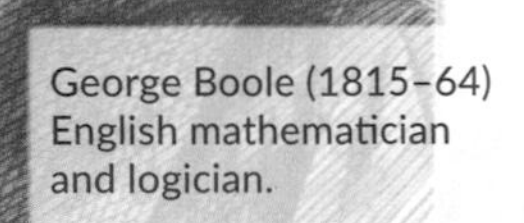

George Boole (1815–64) English mathematician and logician.

1/ Helicopter view: Sometimes it helps to ask what would happen if a thing we suspect is true (but cannot prove yet) were in fact false. The consequences, when we tease them out, may lead us to a contradiction.

In classical **logic**, every factual statement is either true or false, even if we do not know which yet; it cannot be both or neither. What's more, if it is true then so are all its implications (the things that must be true if it is).

Now suppose I have a mathematical statement *S* I want to prove, but do not know how to do it directly. One way is to pretend for a moment that it is false and look at the implications of that. If I find one that contradicts what we already know, I can conclude that *S* cannot be false. And since *S* must be either true or false, I can conclude that it is true.

2/ Shortcut: Imagine I want to prove that every even **natural number** is followed by an odd whole number. Well, what would happen if that turned out to be false? That would mean there was some even number n such that $n + 1$ is also even.

Now we look at the implications. If n is even, it can be written as $2m$ for some other number m. Then $n + 1 = 2m + 1$. But that isn't even! It leaves a remainder of 1 when divided by 2. So every even number *is* followed by an odd number after all!

See also //

5 Logic, p.14

14 Natural Numbers, p.32

27 Cantor's Diagonalization Argument, p.58

54 Pathological Functions, p.112

69 Impossible Constructions, p.142

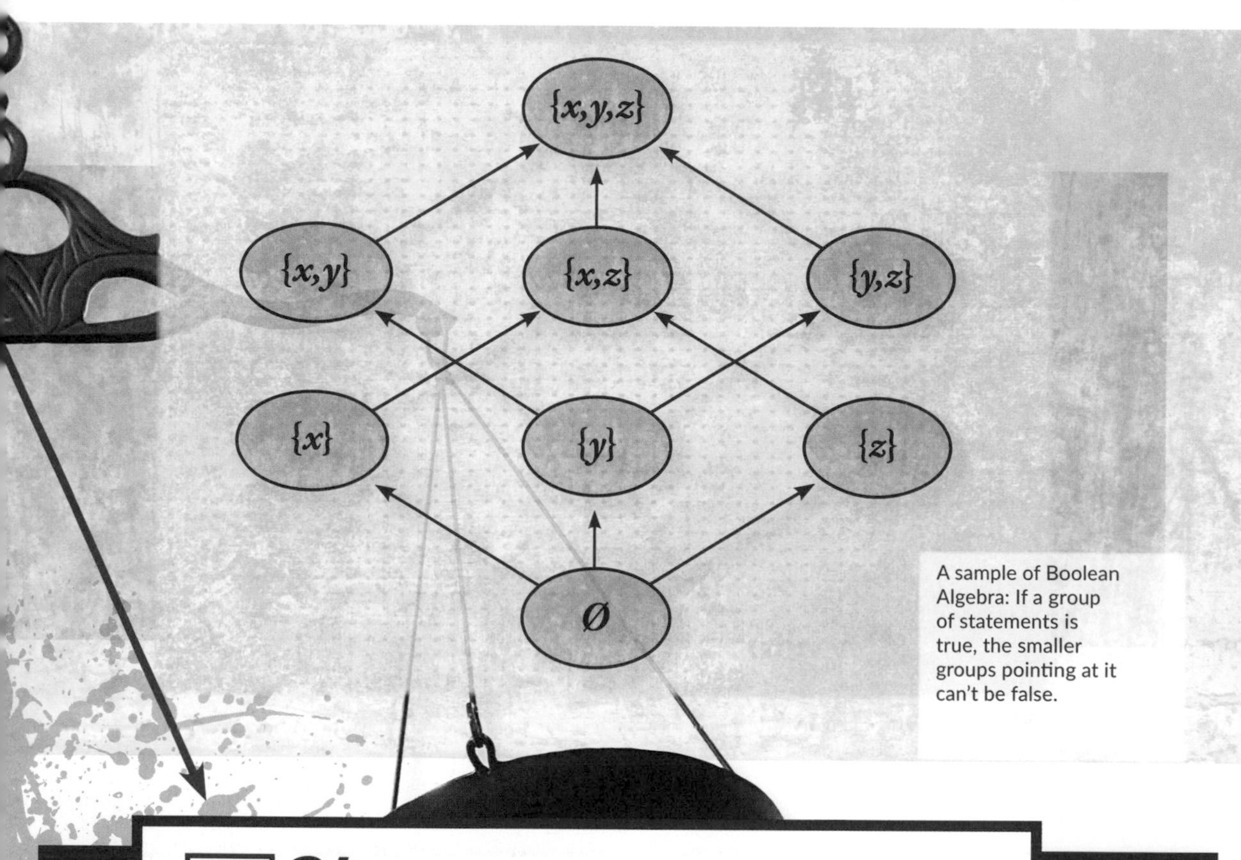

A sample of Boolean Algebra: If a group of statements is true, the smaller groups pointing at it can't be false.

3/ Hack: Every statement is either true or false. If a statement is true so are all its logical consequences. Contradictions are never true.

You can prove something is *true* by temporarily assuming it's *false* and showing that leads to a contradiction.

No.4

Limits Closer and closer . . .

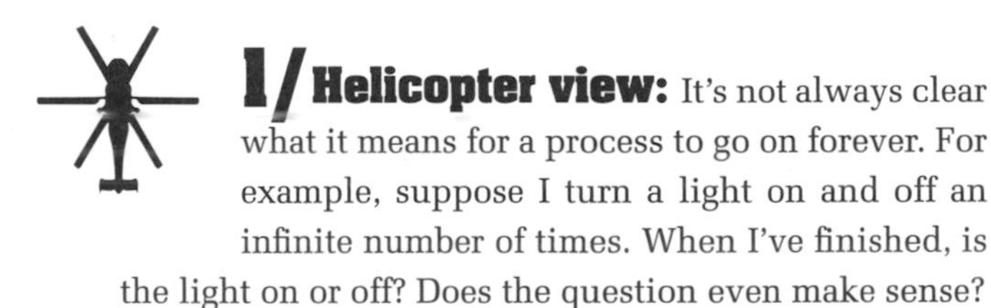

1/ Helicopter view:

It's not always clear what it means for a process to go on forever. For example, suppose I turn a light on and off an infinite number of times. When I've finished, is the light on or off? Does the question even make sense?

Problems like these besieged mathematicians who were trying to make sense of some of the weirder implications of **calculus** around 1800. Some infinite processes were obviously illegitimate, but others seemed to make sense and provide true and useful results.

Limits provide a way to characterize the end state of certain infinite processes, specifically, those that get closer and closer to some value without necessarily ever arriving.

If an infinite process does this, we can say that it is equal to the value that it approaches "in the limit" without actually carrying out an infinite number of steps to get there.

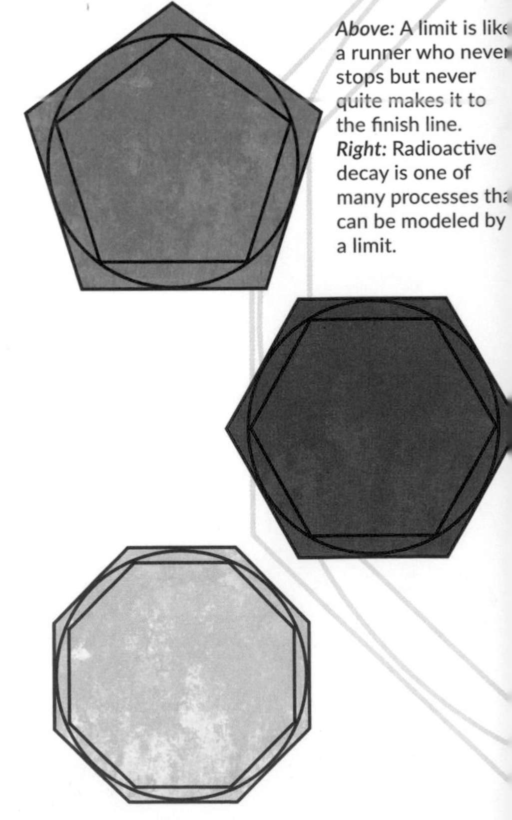

Above: A limit is lik a runner who neve stops but never quite makes it to the finish line. *Right:* Radioactive decay is one of many processes th can be modeled by a limit.

2/ Shortcut:

A lump of uranium-238 loses about half its radioactivity every 4.5 billion years. Imagine we can measure how radioactive it is with a single, perhaps fractional, number.

Every time you check, it will still have some radioactivity; since the number just keeps being divided in half it can never quite make it to zero.

But it gets as close as you like to zero if you wait long enough, so we call zero the *limit* of the process.

See also //

26 Real Numbers, p.56

53 The Fundamental Theorem of Calculus, p.110

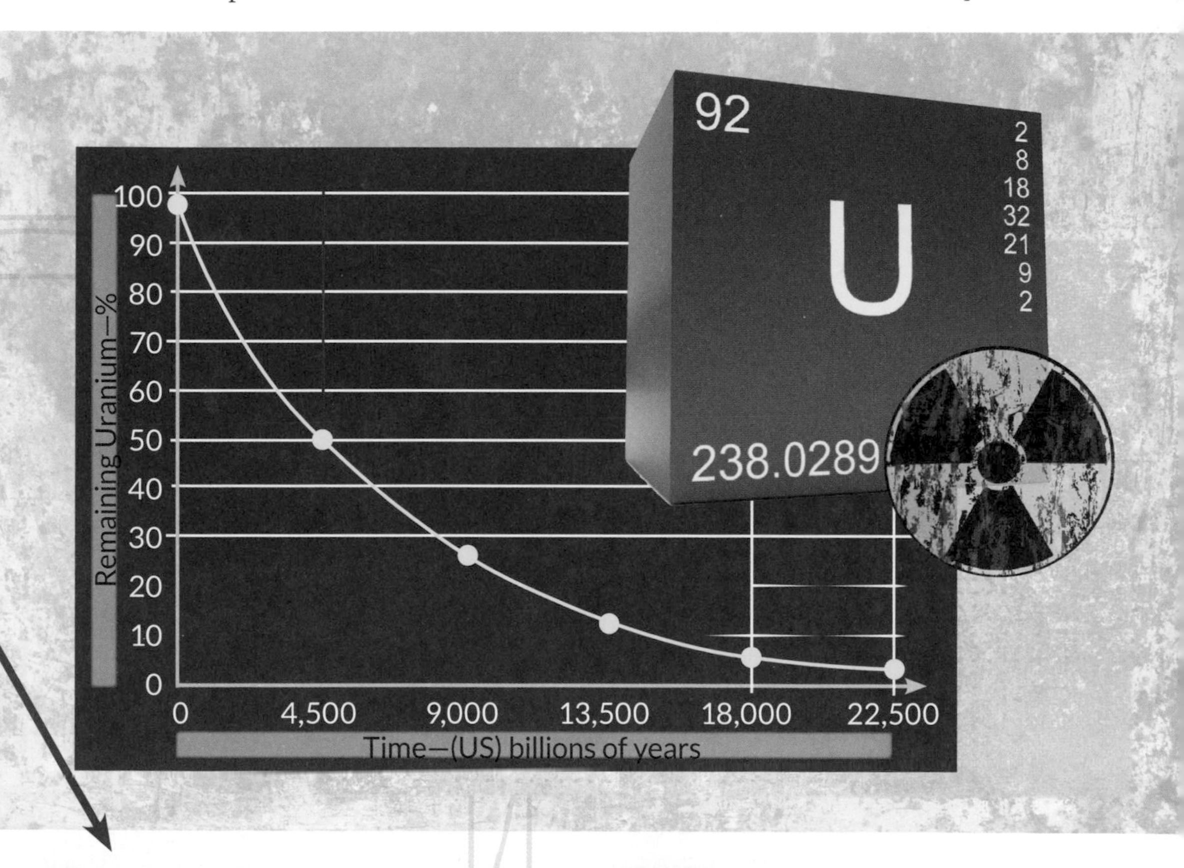

3/ Hack:

Infinite processes sometimes gravitate ever closer to some state without ever getting there, so the limit is that state.

A process gets closer and closer to its limit, if it has one.

No.5
Logic The laws of thought

1/Helicopter view: Aristotle made some of the earliest studies in the proper ways to draw conclusions from available information. Crucially, he turned our attention away from the content of an argument toward its form. For centuries, Aristotle's analysis was the academic gold standard in both the Christian and Islamic worlds.

Most early discussions of logic focused on ordinary language, since that's what most arguments are made of, although symbols were often used to simplify things. In the 19th century, though, logicians began to look at logic from a purely symbolic standpoint.

Logic is part of the foundation of mathematics because of the importance of proofs. Once formalized, though, it also became an object of mathematical research that was studied in its own right. Here's another surprise. This seemingly abstract, purely theoretical subject found a very important practical application in the invention of the computer.

Right: Logic gates are the foundation of modern digital computers.

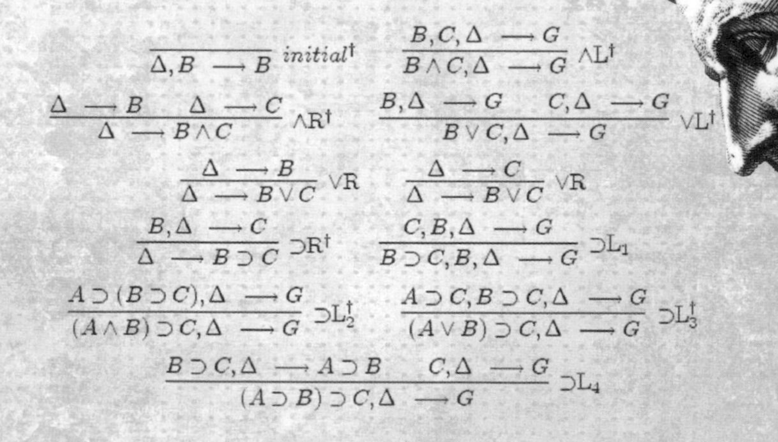

Aristotle (384–322 B.C.E.) was the first to lay down detailed rules for logical validity.

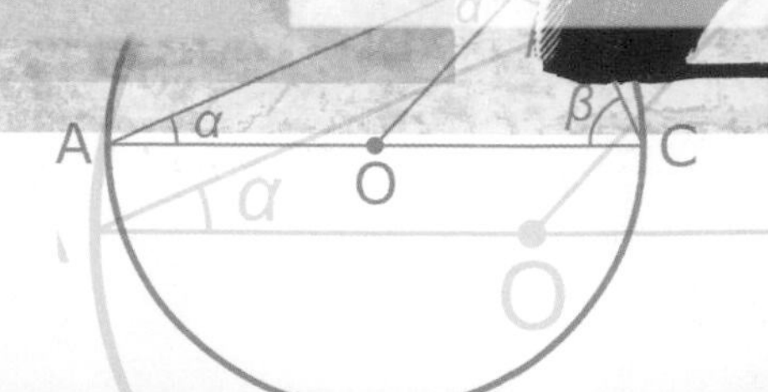

2/ Shortcut: An argument begins with some information, often called assumptions or **axioms**, and draws conclusions from them using **proofs**. If the argument is good, it shouldn't lead us from true assumptions to false conclusions. (If our assumptions are false, though, all bets are off.)

If the logical form of an argument is correct, it is *valid*. A valid argument should never lead us from truth to falsehood. Once you have a good method for turning arguments into symbols, testing their validity is a purely mechanical process.

See also //

1 Axiom, Theorem, Proof, p.6

6 Gödel's Incompleteness Theorems, p.16

35 Abstract Algebra, p.74

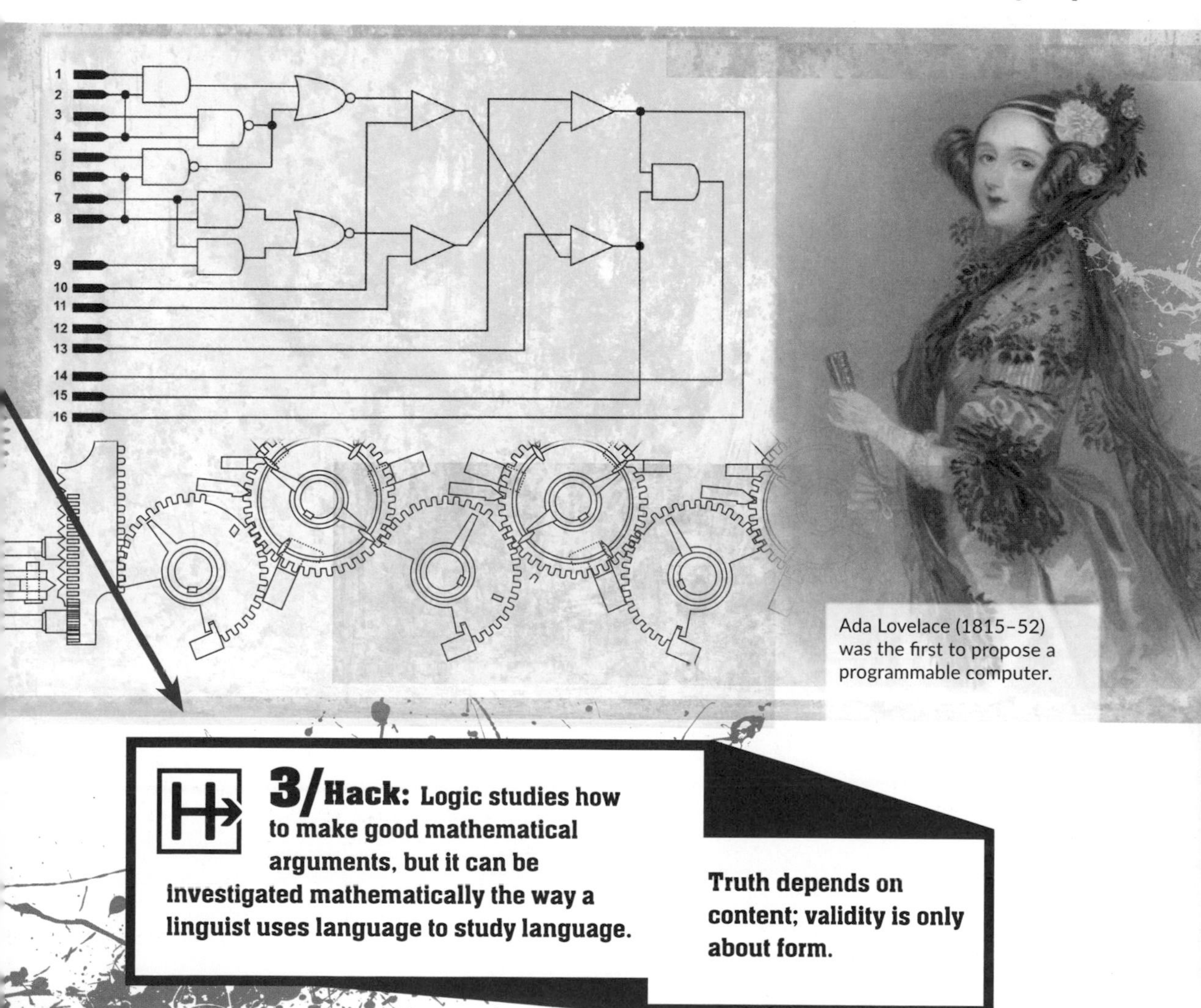

Ada Lovelace (1815–52) was the first to propose a programmable computer.

3/ Hack: Logic studies how to make good mathematical arguments, but it can be investigated mathematically the way a linguist uses language to study language.

Truth depends on content; validity is only about form.

No.6 Gödel's Incompleteness Theorems

Knowing what we cannot prove

Kurt Gödel (1906–78) made many advances in mathematical logic.

1/Helicopter view: A formal theory can be expressed purely in symbolic **logic**. Its **axioms** are known, and every theorem has a logically valid proof. These proofs can be checked mechanically (even by a computer), eliminating mistakes and ambiguities.

For any statement it can make, such a theory should allow us to construct either a **proof** or a disproof. If it can do both, we say the theory is not *consistent*—it proves contradictory things and is not mathematically interesting. If it can do neither, the theory is *incomplete*, which makes it too weak to answer all the questions we expect it to.

In the early 20th century many mathematicians and philosophers hoped that all of mathematics could be reduced to consistent and complete formal theories. Gödel's two Incompleteness Theorems dashed those hopes, proving that consistent, interesting mathematical theories are often doomed to be incomplete.

2/ Shortcut: Gödel's First Theorem says that there are statements about the **natural numbers** that are true, but that are unprovable within the system.

Gödel's Second Theorem says that such a theory can never prove its own consistency. One way to prove a theory is inconsistent is to find a proof within the theory of its own consistency!

Note that the proofs depend on arithmetic structures and do not, contrary to popular belief, apply to other fields of human knowledge.

See also //
1 Axiom, Theorem, Proof, p.6
5 Logic, p.14
29 The Continuum Hypothesis, p.62
44 Diophantine Equations, p.92

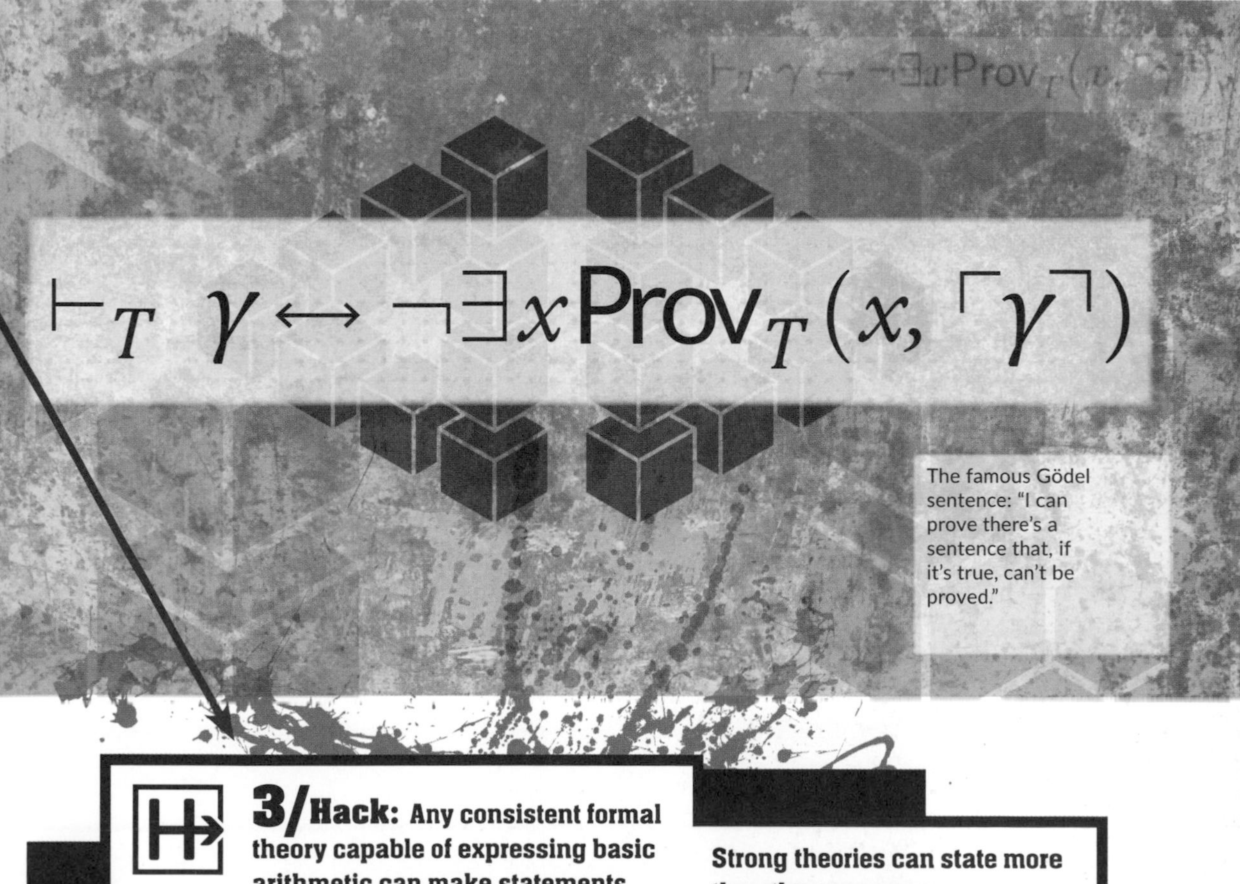

The famous Gödel sentence: "I can prove there's a sentence that, if it's true, can't be proved."

3/ Hack: Any consistent formal theory capable of expressing basic arithmetic can make statements it cannot prove or disprove, and cannot prove its consistency.

Strong theories can state more than they can prove.

No.7

Georg Cantor (1845–1914) invented set theory and made infinity a respectable object of study.

Set Theory

Simple building blocks

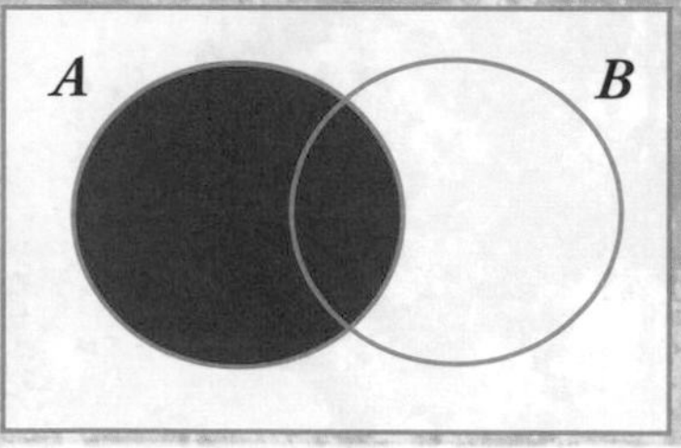

$P(A)$

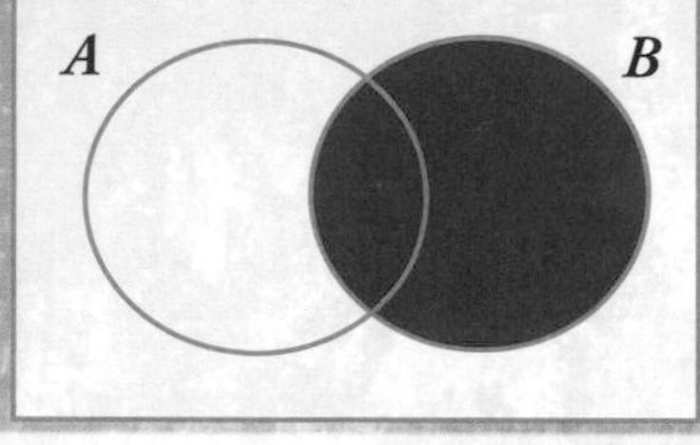

$P(B)$

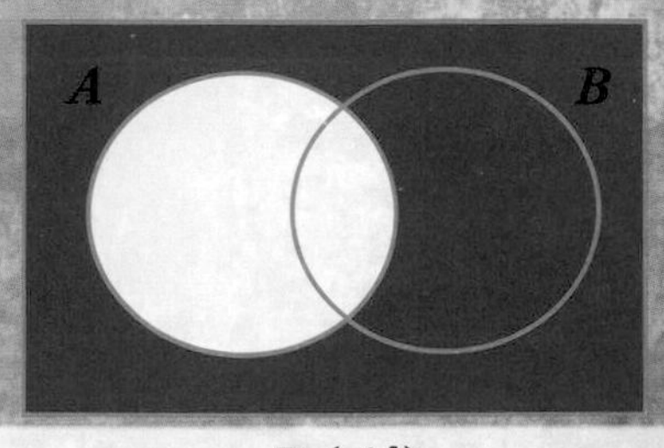

$P(A')$

1/ Helicopter view: Some time around 1874, mathematician Georg Cantor began to use the simple notion of collections of objects as the basis of his investigation of *infinity*. The objects in these collections were not usually specified because, in a sense, they could be anything at all. Such a simple concept allowed questions about numbers to be asked very abstractly, without what we already know about *actual* numbers getting in the way.

Ten years later Gottlob Frege attempted to use *sets*—as the collections were called—as a foundation for all of mathematics, but Bertrand Russell discovered a contradiction in his work. This resulted from ambiguities in the intuitive ideas that Cantor and Frege had been working with.

In the early 20th century these ideas were formalized. Ernst Zermelo and Abraham Fraenkel devised what is now the best-known rigorous version of set theory. This now serves as an almost universal tool in mathematics, providing ways to explicitly make complex objects from simple parts.

Above: Sets provide a simple and intuitive language for almost every part of math.

2/Shortcut: Given two sets A and B, we can get new sets in several ways. Their *union* is the set containing everything in A or in B or in both. Their *intersection* is the set containing only those *elements* that are in both A and B. The set A minus B contains everything in A that is not in B.

Set theory is like a set of building blocks that can be put together in ingenious combinations. Many mathematical ideas can be modeled with sets, making them easier to describe and investigate.

See also //
5 Logic, p.14
13 Categories, p.30
29 The Continuum Hypothesis, p.62
35 Abstract Algebra, p.74

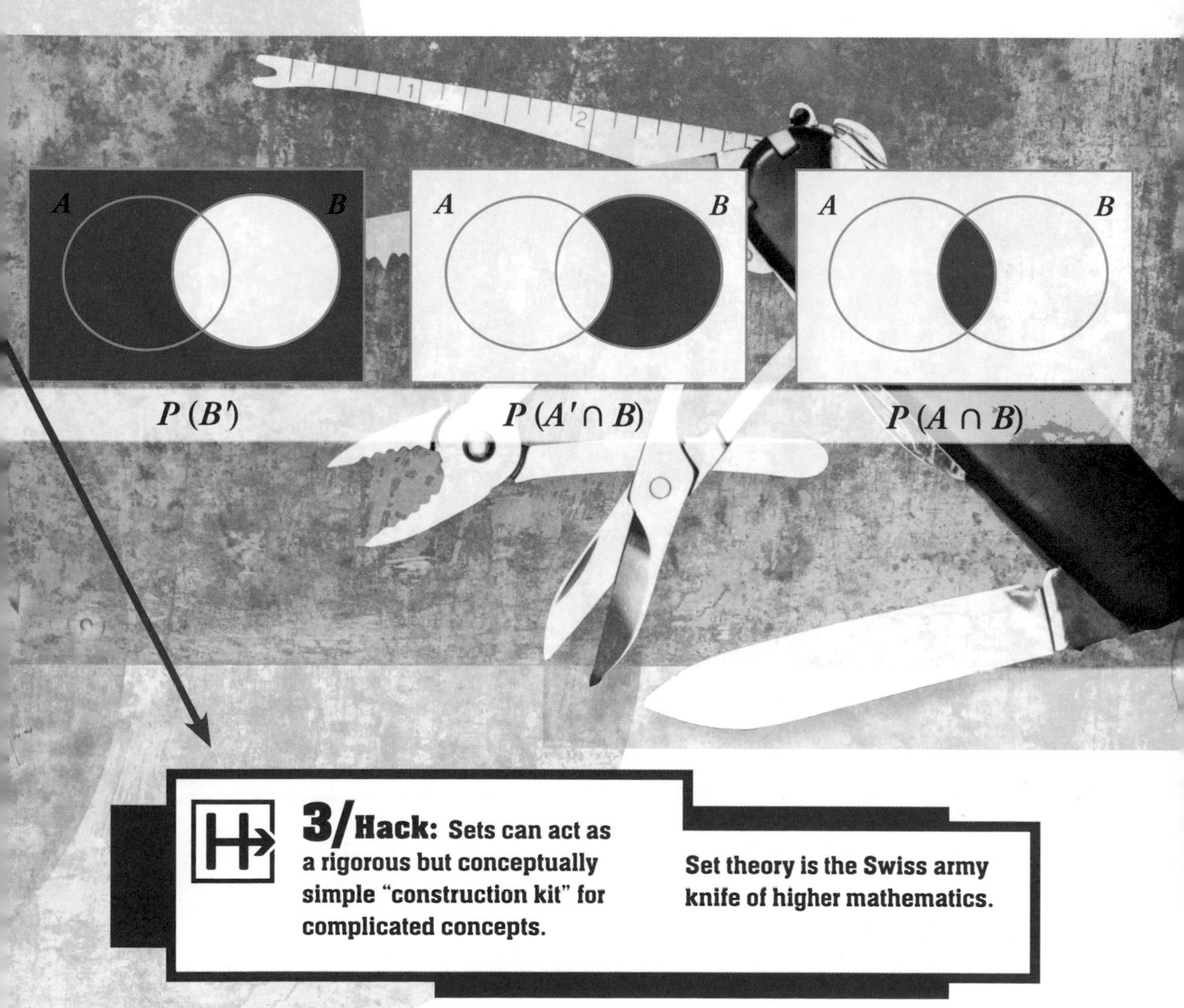

3/Hack: **Sets can act as a rigorous but conceptually simple "construction kit" for complicated concepts.**

Set theory is the Swiss army knife of higher mathematics.

No.8
Products Multiply more things

1/ Helicopter view: A chessboard has eight rows, each containing eight squares, one from each column. That means there are 8 × 8 = 64 squares in total. Each square can be identified by a unique pairing of a row with a column, and squares are what the chessboard is made of. In a sense, the chessboard is the product of its rows and columns.

Once **set theory** came into regular use, it became apparent that this same pattern is repeated in many other mathematical objects. For example, the two-dimensional (2D) space of school geometry can be thought of as the product of a line with another line; and the 3D space we live in is just that 2D space "multiplied by" the same line again.

Such products generalize the operation of multiplying from elementary arithmetic, making it possible to multiply almost any mathematical objects you can think of.

Below: A square can be seen as a product of a line with itself; to make a cube, multiply by a line again.
Right: The Cartesian product of *A* and *B* is the set of all the pairs you can make with one element from each.

2/Shortcut: The product of two sets, A and B, is a new set $A \times B$. Every element of this set is an ordered pair of elements, one from A and one from B. So if A is the set of all appetizers at a restaurant and B the set of all main courses, $A \times B$ contains all the possible two-course meals you can order.

When the two sets share some additional structure their product can often be made to inherit it in a natural way, leading to products of groups, fields, vector spaces, and so on.

See also //
7 Set Theory, p.18
35 Abstract Algebra, p.74
59 Euclidean Spaces, p.122

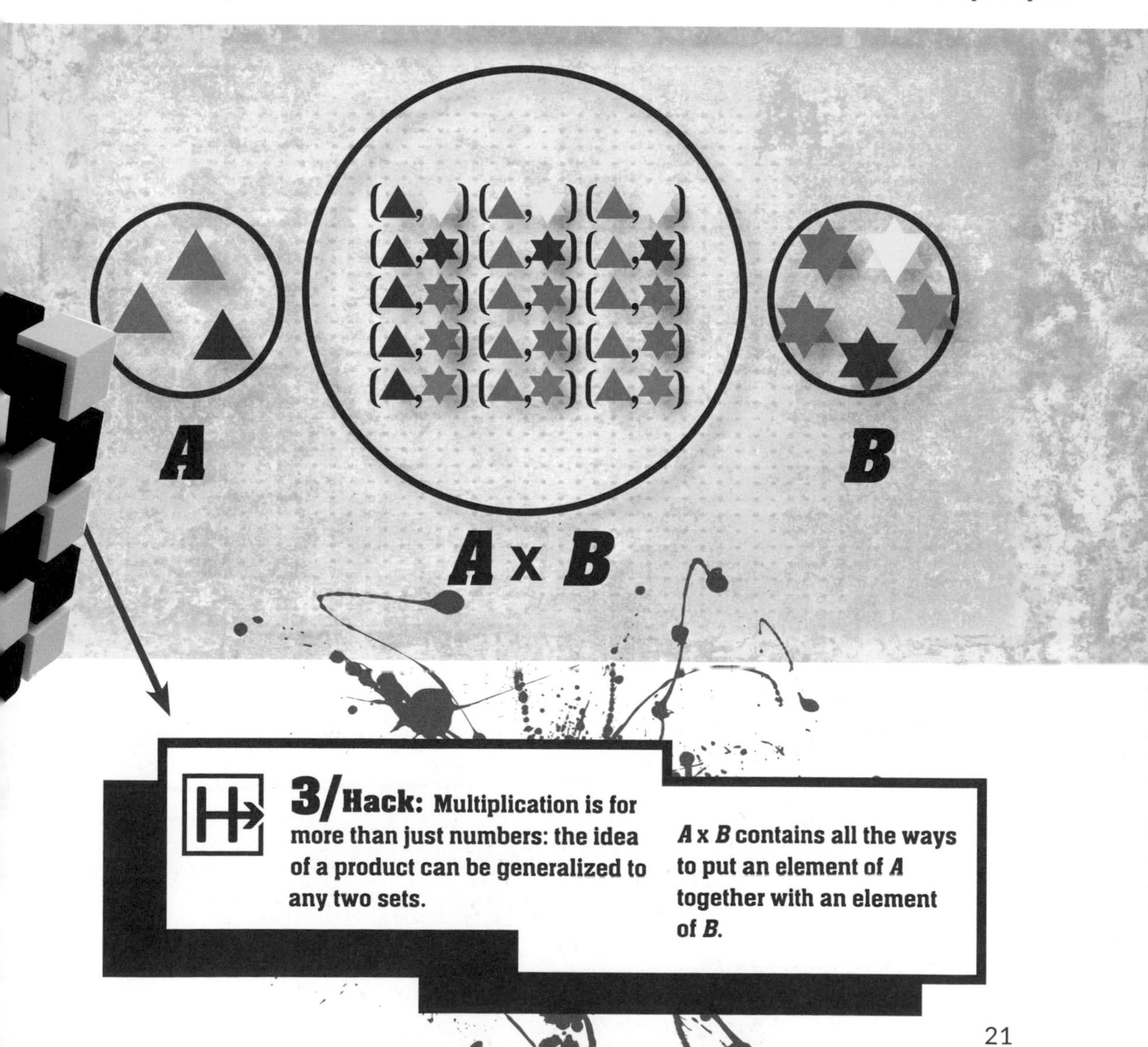

3/Hack: Multiplication is for more than just numbers: the idea of a product can be generalized to any two sets.

$A \times B$ contains all the ways to put an element of A together with an element of B.

No.9
Maps Only connect

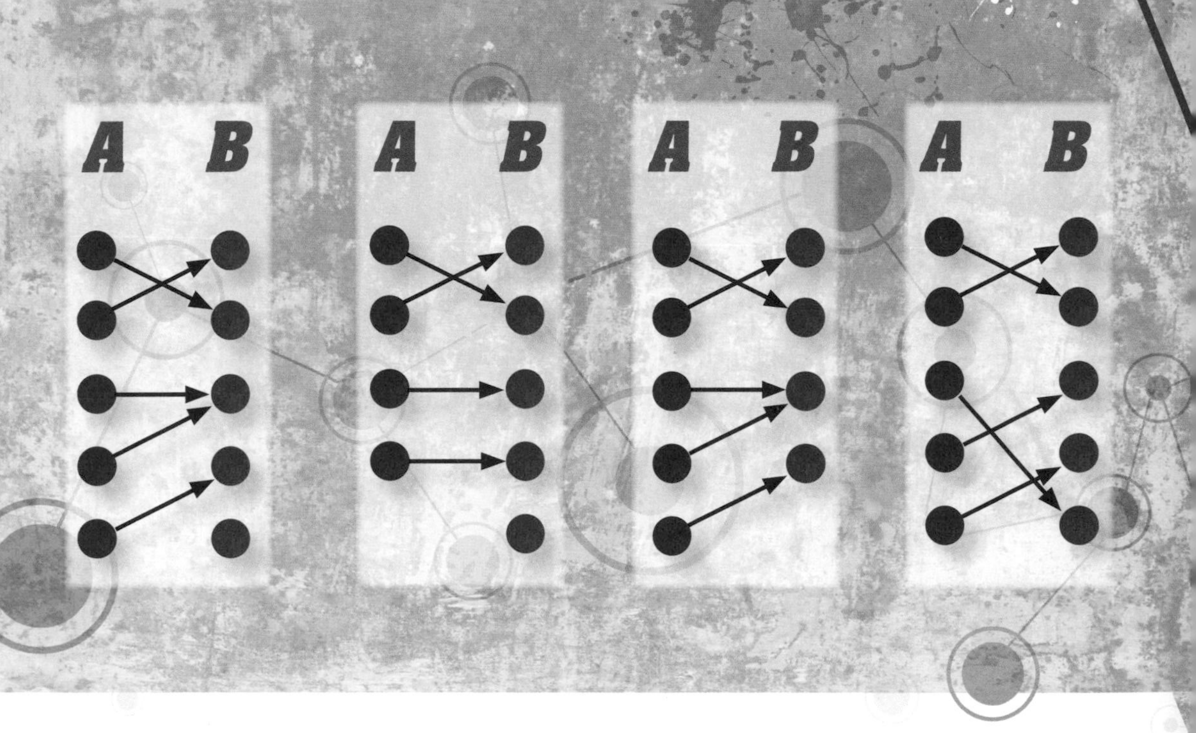

1/ Helicopter view: Sets become really powerful when we join them together. This is done using *maps*. Start with two sets, A and B. Then, a map from A to B takes each element of A and associates an element of B with it.

You can represent a map in a diagram with the elements of A clustered on one side, and those of B on the other, drawing an arrow from each element of A to some element of B.

The set the arrows come from is called the domain of the map, and the set the arrows go to is the codomain.

The rules are that every element of the domain must point to something in the codomain and no element can point to more than one thing. For the codomain, however, there are no rules, and it is acceptable to have multiple arrows pointing at one element and none at all pointing at another.

Above: A map from A to B must have exactly one arrow coming from each element of A and landing somewhere in B.
Right: The figures from Leonardo's *The Last Supper* (*c.*1498) mapped onto their names in the Gospel of John.

2/ Shortcut:

Here is an everyday example. Suppose A is the set of people at a restaurant table and B is the set of meals on the menu. When the diners give their order, they describe a map from A to B.

The two rules then mean that every diner must order something, and nobody is allowed to order two meals. If either of these is broken, we do not have a valid map. On the other hand, we do not require that everyone orders different dishes, or that someone orders every dish!

See also //
7 Set Theory, p.18
13 Categories, p.30

3/ Hack:

A map sends each of the elements of one set to an element of another.

Every element of the domain fires one arrow into the codomain.

No.10 Equivalence Divide and conquer

1/ Helicopter view: Equivalence is the grown-up version of equality. It provides a way of "zooming out" from objects that contain more detail than we care about.

For example, suppose you have a set of shapes made of straight lines. You might decide to group them according to how many sides they have, such as the triangles, quadrilaterals, pentagons, and so on. For the purposes of the grouping we say two shapes are *equivalent* if they have the same number of sides. If S and T are shapes, we might write $S \approx T$ if S and T have the same number of sides, although they might be very different in other ways.

This *partitions* the original set into subsets that represent a sort of classification. In this case the set of triangles, the set of quadrilaterals, and so on. Every shape in the original set is in one of these subsets, and no shape is in more than one. We call these subsets equivalence classes.

These soccer players are grouped into equivalence classes corresponding to their teams.

2/ Shortcut:

A *binary relation* "≈" on a set S is defined as follows: if a and b are elements of S, then $a \approx b$ is either true or false. For example, on the set of counting numbers we have the binary relation "<"; you already know that $3 < 5$ is true and $16 < 2$ is false.

An *equivalence relation* is a binary relation defined by three extra **axioms**. First, $a \approx a$ is always true. Second, if $a \approx b$ is true, $b \approx a$ must be too. Finally, if $a \approx b$ and $b \approx c$ are true, $a \approx c$ must also be true. Note that "<" is not an equivalence relation!

See also //

1 Axiom, Theorem, Proof, p.6
7 Set Theory, p.18
8 Products, p.20

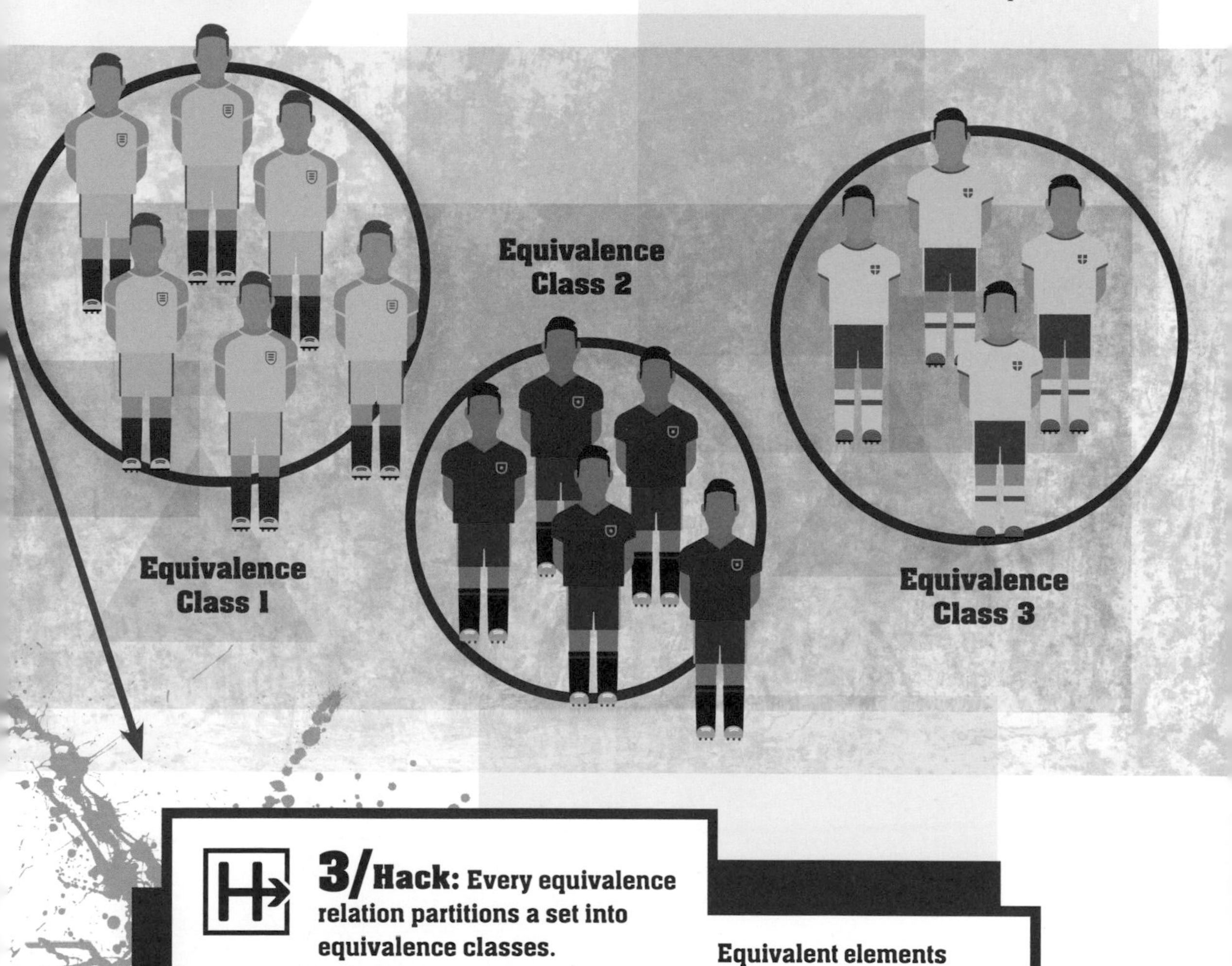

3/ Hack:

Every equivalence relation partitions a set into equivalence classes.

Equivalent elements stick together.

No.11
Inverses Things come undone

1/ Helicopter view: A **map** between sets can be thought of as an action or transformation. Think of traveling along the arrows from one side of the diagram to the other. The question is, can we use another map to get back again? If we can, the map has an *inverse*, and inverses are of central importance in mathematics.

Many mathematical operations can—at least sometimes—be reversed. Subtraction undoes addition, division undoes multiplication, and so on. This theme continues in more abstract mathematics, too.

We undo a map by turning the arrows around. This swaps the roles of domain and codomain. But we can only do this under certain circumstances.

A map from diners to menu items might have no inverse because two people ordered the same dish, or because nobody ordered one of the dishes. Either way, turning the arrows around does not give us a valid map, so there is no inverse.

Right: A function f and its inverse f^{-1}. Because f maps a to 3, the inverse f^{-1} maps 3 back to a.

2/Shortcut: If I is a map from set A to set B, its inverse is a map from B to A. It's just like I but the arrows point the opposite way.

It only exists if I points a single arrow at each element in B, no more and no less. The technical term for this type of map is a *bijection*.

See also //
7 Set Theory, p.18
9 Maps, p.22
38 Groups, p.80
42 Rings and Fields, p.88

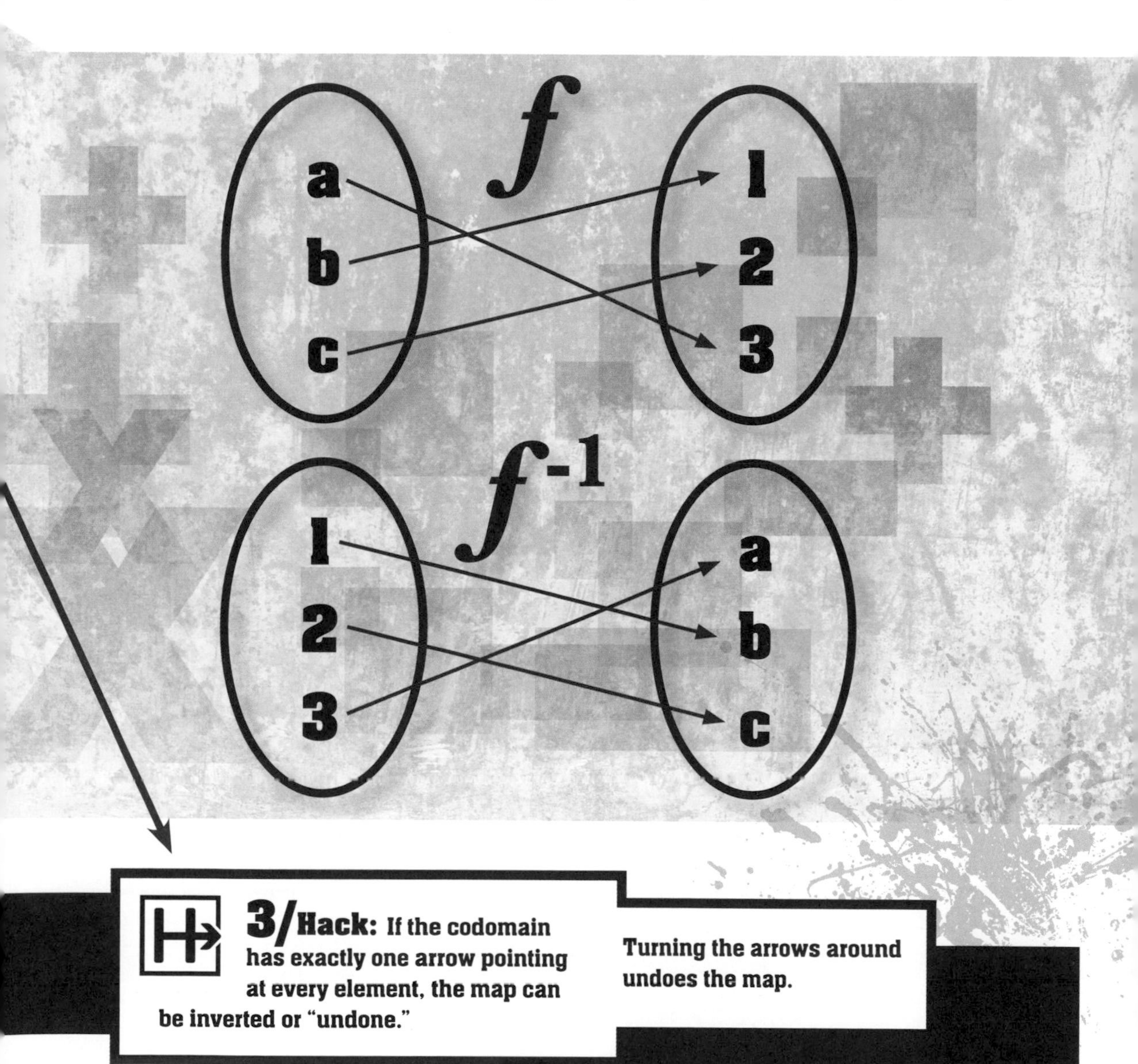

3/Hack: If the codomain has exactly one arrow pointing at every element, the map can be inverted or "undone."

Turning the arrows around undoes the map.

No.12 The Schröder–Bernstein Theorem

Establishing equality

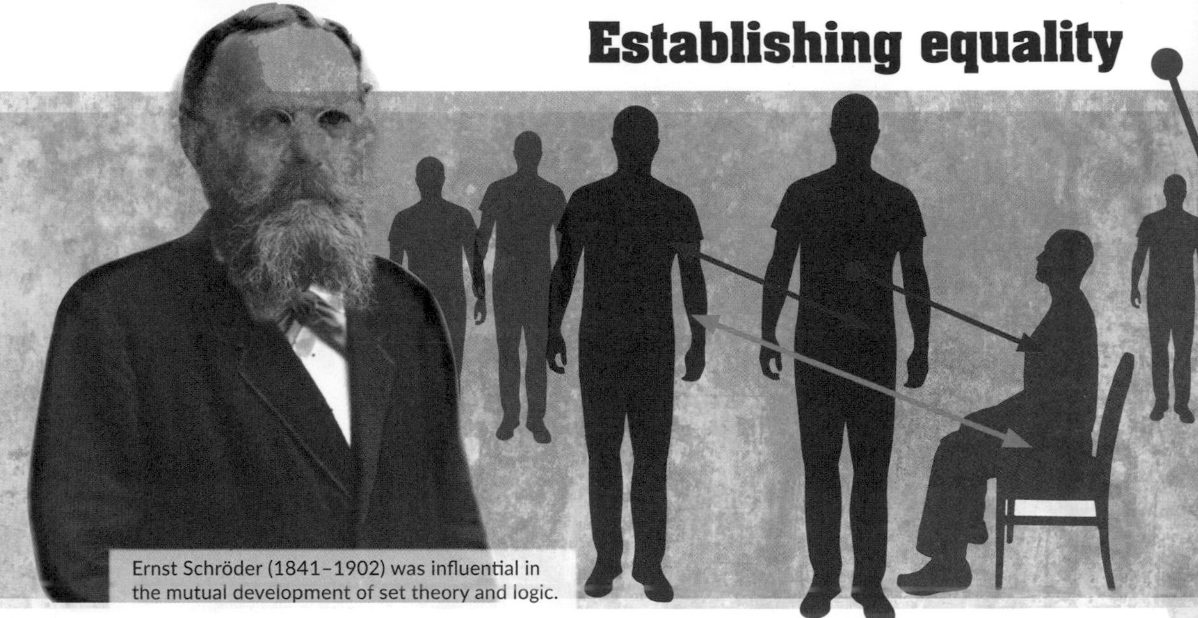

Ernst Schröder (1841–1902) was influential in the mutual development of set theory and logic.

1/ Helicopter view: Counting is surprisingly hard in mathematics, especially when infinite numbers are involved. **Set theory** made it possible for the first time to prove things about infinite sets in a rigorous way. The *Schröder–Bernstein Theorem* is an extremely useful example.

Imagine a classroom with a number of chairs and some students. If each chair has a student sitting in it, we can conclude there are at least as many students as chairs (there might be more, as some students might be standing up). If every student has a chair, on the other hand, we conclude there are at least as many chairs as there are students (there might be some unused chairs).

Now imagine both of these are true. Then it must be that the numbers of chairs and students are exactly the same. This seems obvious, but it becomes more powerful if we extend it to infinite sets (such as some sets of numbers). This generalization is what the Schröder–Bernstein Theorem achieves.

Right: We don't have to count to see whether there are exactly as many students as chairs.

2/ Shortcut: Many proofs use Schröder–Bernstein to show that two sets—let us call them A and B—have the same number of elements. They work by constructing two **maps**. One goes $A \to B$ so that no element of B has more than one arrow pointing at it. The other goes $B \to A$ with at most one arrow pointing at each element of A.

If such **maps** exist, we get an immediate proof that A and B have the same number of elements, even if they are infinite.

See also //
7 Set Theory, p.18
9 Maps, p.22
11 Inverses, p.26
16 Hilbert's Hotel, p.36

3/ Hack: If A is at least as big as B and B is at least as big as A, they must be the same size.

A pair of maps can be used to compare the sizes of two sets.

No.13 Categories Generalized nonsense

1/Helicopter view: Mathematics often proceeds by abstraction and generalization. It can be remarkably fruitful simply to notice a pattern you have seen in several places, separate it from those specific contexts, and then look for other places where the same pattern occurs. This generalized understanding can then be applied to wildly different problems.

Category theory has at its heart the idea that mathematics, and how mathematicians think about it, is a patterned activity. In particular, it aims to understand the patterns of thought that lie behind apparent flashes of genius or insight that lead to breakthroughs. One of its goals is to help you—a mere mortal, presumably—to spot the same patterns that a great mathematician might.

In its early days, category theory was derided as "abstract, generalized nonsense" or "comic-book mathematics," a reference to its preference for diagrams over symbolic or verbal arguments. But its success in solving real problems has since silenced such criticisms.

Below: Exact sequences (left–right) and natural transformations (up–down) are key ingredients of category theory. *Right:* Abstracting the general form of an idea from the context allows it to be applied in other contexts more easily.

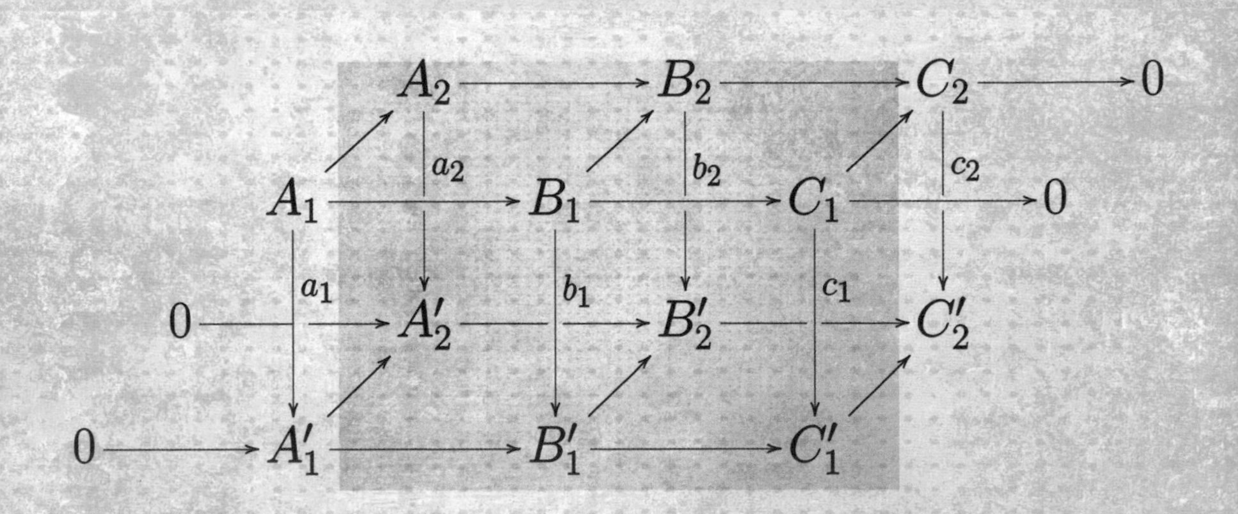

2/Shortcut: A *category* consists of two things; a collection of objects and a collection of mappings between them, called *morphisms*, governed by a few simple rules. The objects might be sets, or they might not. Similarly, the *morphisms* might or might not be **maps**.

The real power of the approach comes from *functors*, which are ways of associating one category's objects and morphisms with those of another. They can help transfer knowledge from one field of mathematics to another, yielding rapid and powerful new results.

See also //
7 Set Theory, p.18
9 Maps, p.22
35 Abstract Algebra, p.74
84 Algebraic Topology, p.172
88 Hilbert's Nullstellensatz, p.180

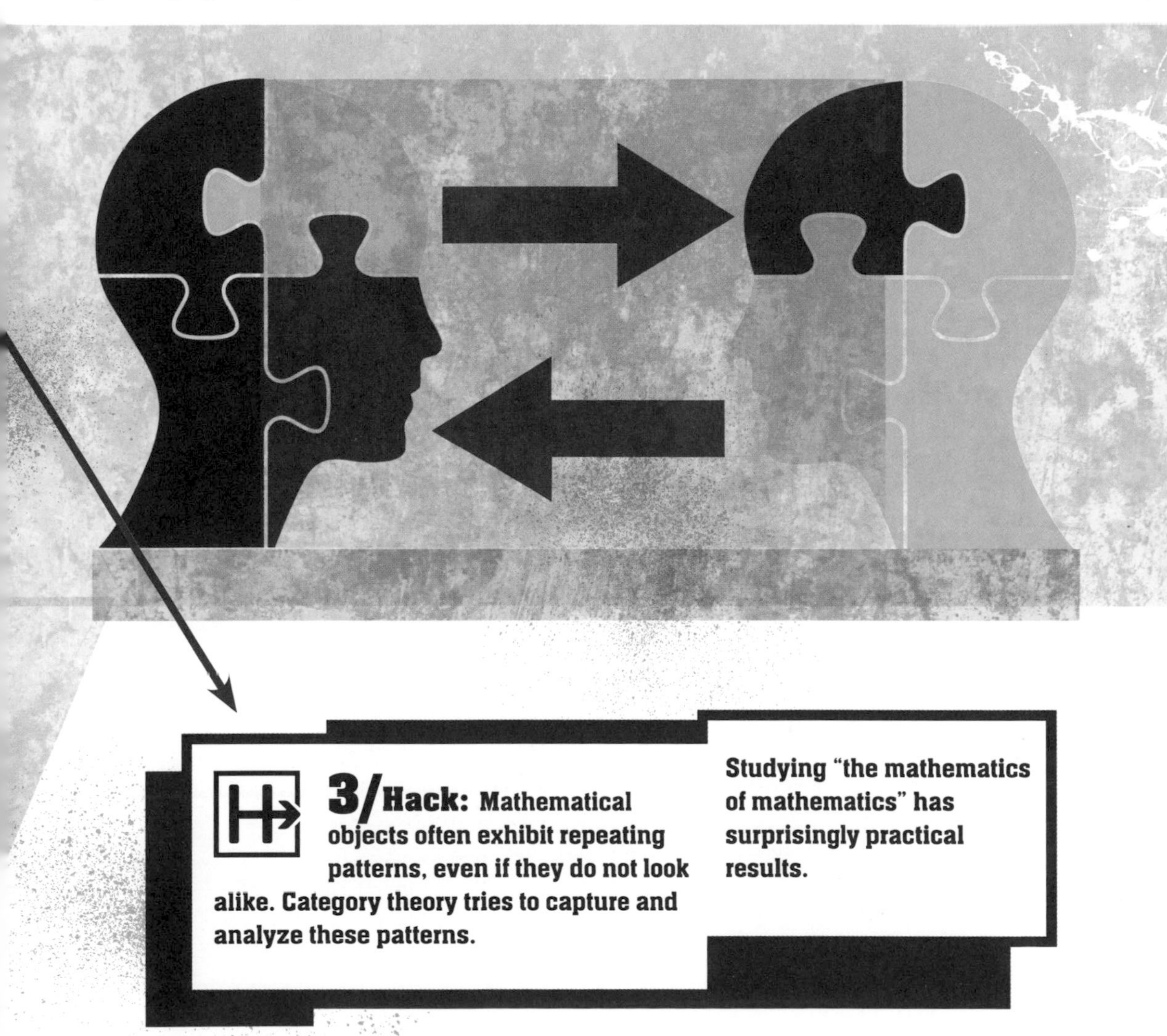

3/Hack: Mathematical objects often exhibit repeating patterns, even if they do not look alike. Category theory tries to capture and analyze these patterns.

Studying "the mathematics of mathematics" has surprisingly practical results.

No.14

Giuseppe Peano (1858–1932) was the first to axiomatize ordinary arithmetic in modern logical language.

Natural Numbers
They're what counts

1/ Helicopter view: Every child learns to count with the ordinary *natural numbers* (or whole numbers). These start 1, 2, 3, 4, and can go as high as we like by repeatedly adding 1 (whether 0 is included is a matter of convention). They are an infinite set and there is no greatest natural number. This, perhaps, is a first clue to their hidden depths.

The natural numbers are the oldest and least exotic of mathematical objects, but they can still raise difficult questions. *Number theory* is the branch of mathematics that studies them. It is one of the few theories in which there are long-standing unsolved problems that can be stated and understood without much technical apparatus.

The natural numbers can be easily added and multiplied, and the result is always another natural number; they are not so good for subtraction and division because the end result may not be a natural number. They can be described by formal **axioms**, such as those devised by Giuseppe Peano (1889). Because of its apparent simplicity, the arithmetic of natural numbers is often used as a logical test case.

Right: The counting numbers are among the first mathematical objects we learn about as children.

2/ Shortcut: The fundamental principle in the theory of the *natural numbers* is **induction**, which gives us the apparently miraculous ability to ascend through their infinite number by a single step.

It works so well because of a principle every child learns, which is that you get from one number to the next by the same process every time, the simple act of adding 1. In Peano's axioms we call this the *successor operation*, and write things like $S(3) = 4$ instead of $3 + 1 = 4$.

See also //
1 Axiom, Theorem, Proof, p.6
2 Induction, p.8
6 Gödel's Incompleteness Theorems, p.16
17 Prime Numbers, p.38
20 Negative Numbers, p.44

3/ Hack: We can add and multiply, not always subtract or divide, but induction makes the natural numbers especially powerful and interesting.

The act of adding 1 is deeper than you may have previously thought.

No.15
The Collatz Conjecture
Easy to state, hard to prove

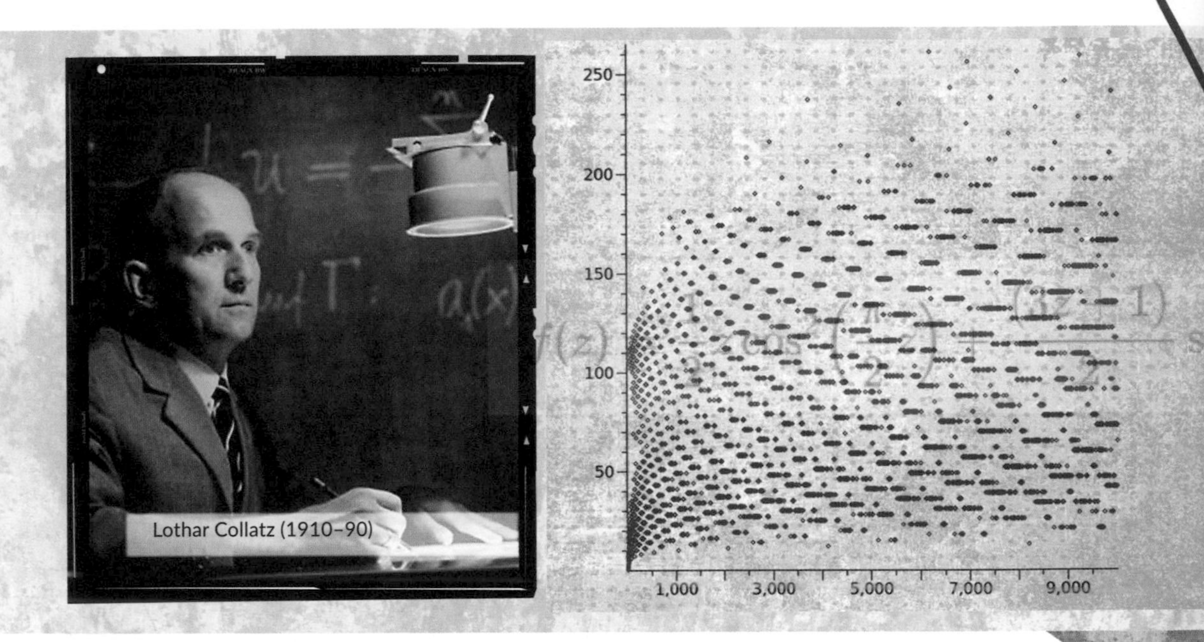

1/ Helicopter view: Here is a game invented by Lothar Collatz at some point in the middle of the 20th century. Pick any **natural number**. If the number is 1, the game is over. If it is even, halve it. Otherwise, multiply it by 3 and add 1. Now take the resulting number and repeat.

Start with 17. This is not even, so triple it and add 1, which gives 52. That is even, so halve it to get 26, halve again to get 13. Again, not even, so triple and add 1; we get 40. Halve it to get 20; halve again to get 10, halve again to get 5. Now triple and add 1 to get 16. Now halve it to get 8, halve again to get 4, halve again to get 2 and again to get 1. The game is now over. The whole sequence was 17→52→26→13→40→20→10→5→16→8→4→2→1. Try it yourself with any number you choose.

Here is the puzzle: Does the game end after a finite number of steps regardless of which number we start with?

Above: Numbers from 1 to 9,999 and their corresponding total stopping time.

2/Shortcut: Collatz's game is *iterative*, that is, it does something to a number and then feeds the result back into itself again and again until it is whittled down to 1. Iterative processes have great importance in technology, especially in computing. But Collatz's process has no known practical applications.

What makes it famous is that nobody has a clue how to solve it. Paul Erdős famously declared it "absolutely hopeless." It is even possible that it is unsolvable because of Gödel's First **Incompleteness Theorem**, but we don't know.

See also //

6 Gödel's Incompleteness Theorems, p.16

14 Natural numbers, p.32

89 Iteration, p.182

91 Chaos Theory, p.186

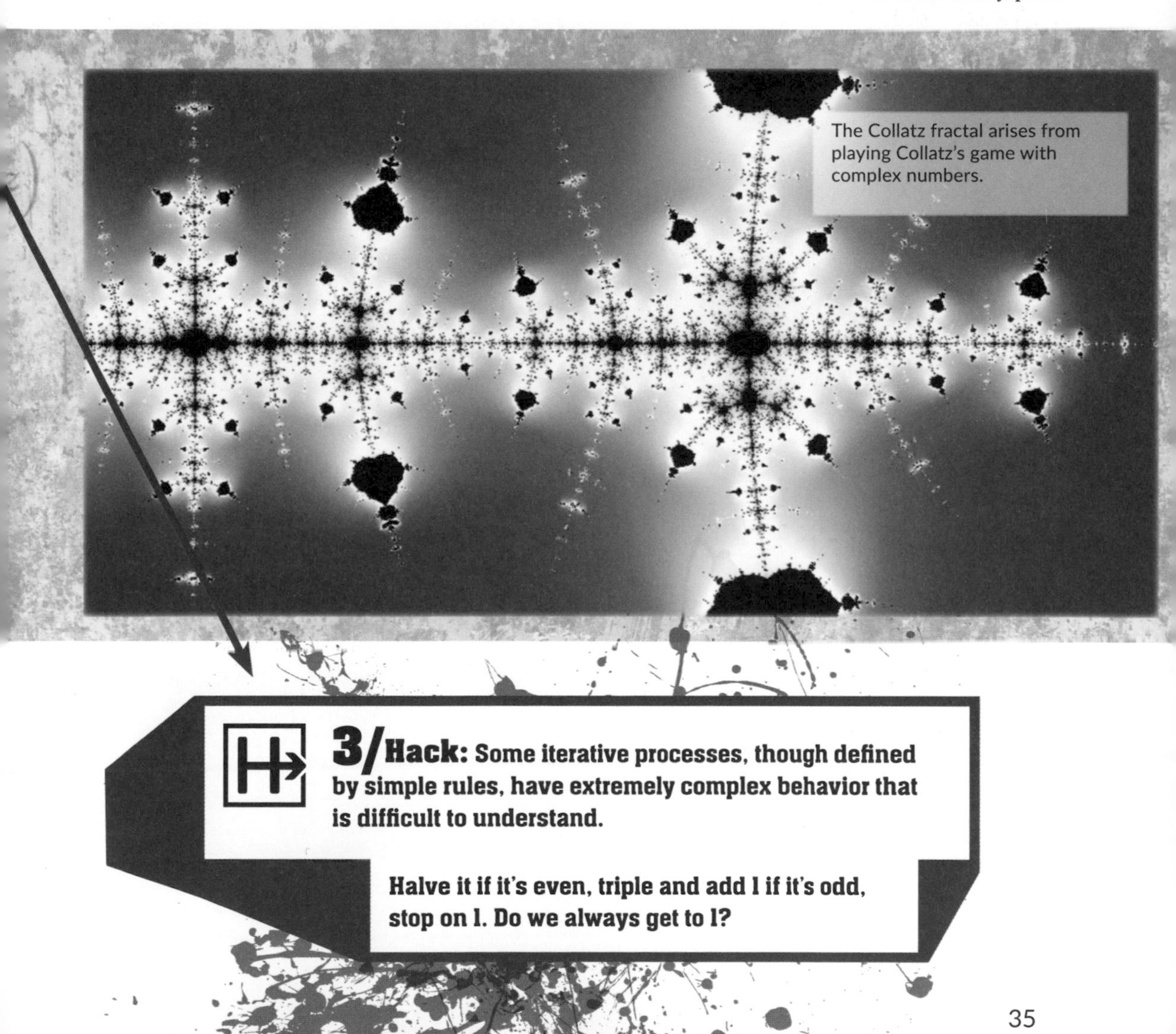

The Collatz fractal arises from playing Collatz's game with complex numbers.

3/Hack: Some iterative processes, though defined by simple rules, have extremely complex behavior that is difficult to understand.

Halve it if it's even, triple and add 1 if it's odd, stop on 1. Do we always get to 1?

No.16
Hilbert's Hotel

The part is equal to the whole

David Hilbert (1862–1943) was one of the most influential mathematicians of his time.

1/ Helicopter view: In 1924 David Hilbert told the following story. Suppose there is a hotel with an infinite number of rooms. Like most hotels, each room is identified by a **natural number**, but in this case every natural number is the number of some room or other. Further suppose that, just now, every room is occupied.

We'd like to fit an extra guest in. How can we do it when every room is occupied? It's quite easy. Ask everyone to move to the room next door, whose number is 1 greater than theirs. Everyone can do this, since every number has a successor. This leaves Room 1 empty for the new guest.

But we can do more. Imagine we now ask everyone to move to the room whose number is double their own. Note, again, that everyone can do this since for every number n, $2 \times n$ is also a number. Now only the even-numbered rooms are occupied, leaving an infinite number of odd-numbered rooms vacant.

Above: Aleph Null refers to the cardinality of a countably infinite set.

2/Shortcut: *Hilbert's Hotel* exhibits a property that only infinite sets can exhibit—a part can seem to be the same size as the whole thing. The size of a set is called its *cardinality*, and infinite cardinalities are one of **set theory**'s most central topics.

Their behavior is so strange that, at least before Cantor, most people believed the infinite was an intrinsically paradoxical idea that had no place in mathematics. Philosophical worries linger, but most mathematicians are now comfortable residents of Hilbert's Hotel.

See also //
5 Logic, p.14
7 Set Theory, p.18
14 Natural Numbers, p.32

3/Hack: **When we consider the natural numbers altogether as a whole, their infinite cardinality has surprising effects.**

An infinite set's parts can be the same size as the whole thing.

No.17 Prime Numbers

The atoms of number theory

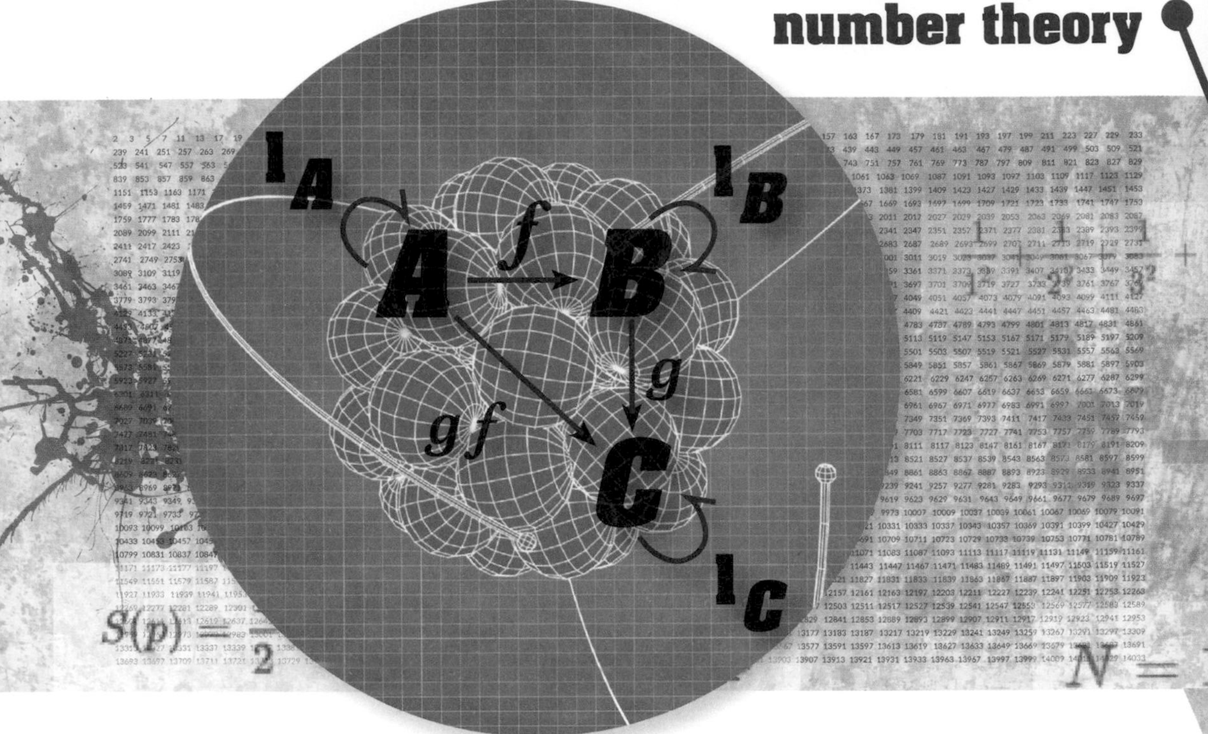

1/ Helicopter view: Many **natural numbers** can be made by multiplying others, for example, 6 is 2 × 3. The process of breaking a number down like this is called factorization, and at some point it comes to a halt.

For example, 20 is 5 × 4 and 4 is 2 × 2, but we cannot break down 5 or 2 any further. We have broken the number 20 down to its smallest parts: 20 is 5 × 2 × 2. These smallest parts are called *prime numbers*, and we can think of them as the building blocks or atoms of the natural number system.

We call 5 × 2 × 2 the *prime factorization* of 20. There is only one possible prime factorization for each number, as Euclid knew back in *c*.300 B.C.E. This fact is so important it is often called the Fundamental Theorem of Arithmetic. The prime factorization is like the fingerprint or chemical formula of a natural number.

Prime numbers have surprisingly complicated properties, and some of the most basic-sounding puzzles about them remain unsolved.

Right: Number theorists think of all whole numbers as being made up of primes multiplied together.

2/ Shortcut: The prime numbers start off as follows: 2, 3, 5, 7, 11, 13, 17, 19, and never run out. However, between 1 and 100 there are 25 primes, but between 101 and 200 there are only 21, and they gradually thin out the higher up we go.

This prompts us to look for a pattern in the distribution of primes within the natural numbers. However, despite a great deal of research we still do not have a complete picture. The simplest elements of the simplest numbers still hold many mysteries.

See also //

14 Natural Numbers, p.32

18 The Twin Prime Conjecture, p.40

19 The Goldbach Conjecture, p.42

Periodic Table of Primes

$$\left(\frac{1}{3}+\frac{1}{5}\right)+\left(\frac{1}{5}+\frac{1}{7}\right)+\left(\frac{1}{11}+\frac{1}{13}\right)+\cdots=\sum_{\substack{p\text{ prime,}\\ p+2\text{ prime}}}\left(\frac{1}{p}+\frac{1}{p+2}\right)$$

2																	3
5	7											11	13	17	19	23	29
31	37											41	43	47	53	59	61
67	71	73	79	83	89	97	101	103	107	109	113	127	131	137	139	149	151
157	163	167	173	179	181	191	193	197	199	211	223	227	229	233	239	241	251
257	263	269	359	367	373	379	383	389	397	401	409	419	421	431	433	439	443
449	457	461	569	571	577	587	593	599	601								

271	277	281	283	293	307	311	313	317	331	337	347	349	353
463	467	479	487	491	499	503	509	521	523	541	547	557	563

3/ Hack: Primes are different from the other natural numbers, and can be seen as the system's basic building blocks.

A prime number has no whole numbers that divide into it (except itself and 1, which do not count).

No.18 The Twin Prime Conjecture

Neighborly numbers

1/ Helicopter view: The numbers 3 and 5 are **prime**, and are as close together as two primes greater than 2 can be; they are consecutive odd numbers. This is also true of 5 and 7, and of 11 and 13, and of 17 and 19. It seems such pairings are, in fact, common and so they are called *twin primes*.

Because primes get rarer as we look at larger and larger numbers, we might be led to think that at some point we just stop getting any more twin primes. The Twin Prime Conjecture says this will not happen and that we can find twin primes as large as we like.

At the time of writing, we don't know whether the Twin Prime Conjecture is true or false, but an international, collaborative project appears to be making good progress. It's possible that it will have been resolved one way or the other by the time you read this.

Above: A color rendition of the sieve of Eratosthenes

2/Shortcut: The way the primes are distributed among all the **natural numbers** remains quite mysterious.

Some very large twin primes have been found by computer. In September 2016, a pair was found that, written in decimal notation, would each fill a fat paperback book.

Nevertheless, this is a drop in the ocean of the infinite natural numbers. To prove the conjecture for all numbers, we must deepen our understanding of primes themselves.

See also //
14 Natural Numbers, p.32
16 Hilbert's Hotel, p.36
17 Prime Numbers, p.38

Terence Tao (b.1975) is associated with an international project attempting to resolve the Twin Prime Conjecture.

3/Hack: A pair of twin primes is as close as primes can get. The conjecture says they never run out.

If *N* is prime and so is *N* + *2*, then they're twins.

No.19 The Goldbach Conjecture

Adding primes

Leonhard Euler (1707–83) has many things named after him thanks to his major contributions to the mathemathics of his day.

Above: The letter from Goldbach to Euler dated June 7, 1742 in which he proposed his conjecture: *Every integer which can be written as the sum of two primes, can also be written as the sum of as many primes as one wishes, until all terms are units.*

1/ Helicopter view: Christian Goldbach wrote to Leonhard Euler in 1742 with a proposition that amounts to this: every even **natural number** greater than 2 can be expressed as the sum of two **primes**. For example, we can write 10 as 5 + 5 or 22 as 3 + 19.

What makes this awkward to prove is that primes are about multiplication whereas this is about addition. If it is true, the proof has stayed hidden for more than 250 years despite great efforts at finding it.

It has even been suggested it might be true, but only by chance. The bigger the number you pick, the more ways there are to represent it as sums of two smaller numbers. You have more primes to choose from, too. These advantages both just keep on growing as the number grows. If, indeed, there is no reason *why* it is true it might turn out to be impossible to prove.

2/Shortcut: Goldbach suggests another way—besides factorization—to break up a **natural number** into **primes**. A prime factorization always exists and is unique, which means we can always do it and everyone who does gets the same result. Existence and uniqueness are two properties that are very nice for a mathematical procedure to have.

Unlike prime factorization, decomposing an even number into the sum of two primes is not always unique. Take an even number N. Are there two primes P and Q so that $N = P + Q$? That's what we still do not know, at least at the time of writing.

See also //

14 Natural Numbers, p.32

17 Prime Numbers, p.38

18 The Twin Prime Conjecture, p.40

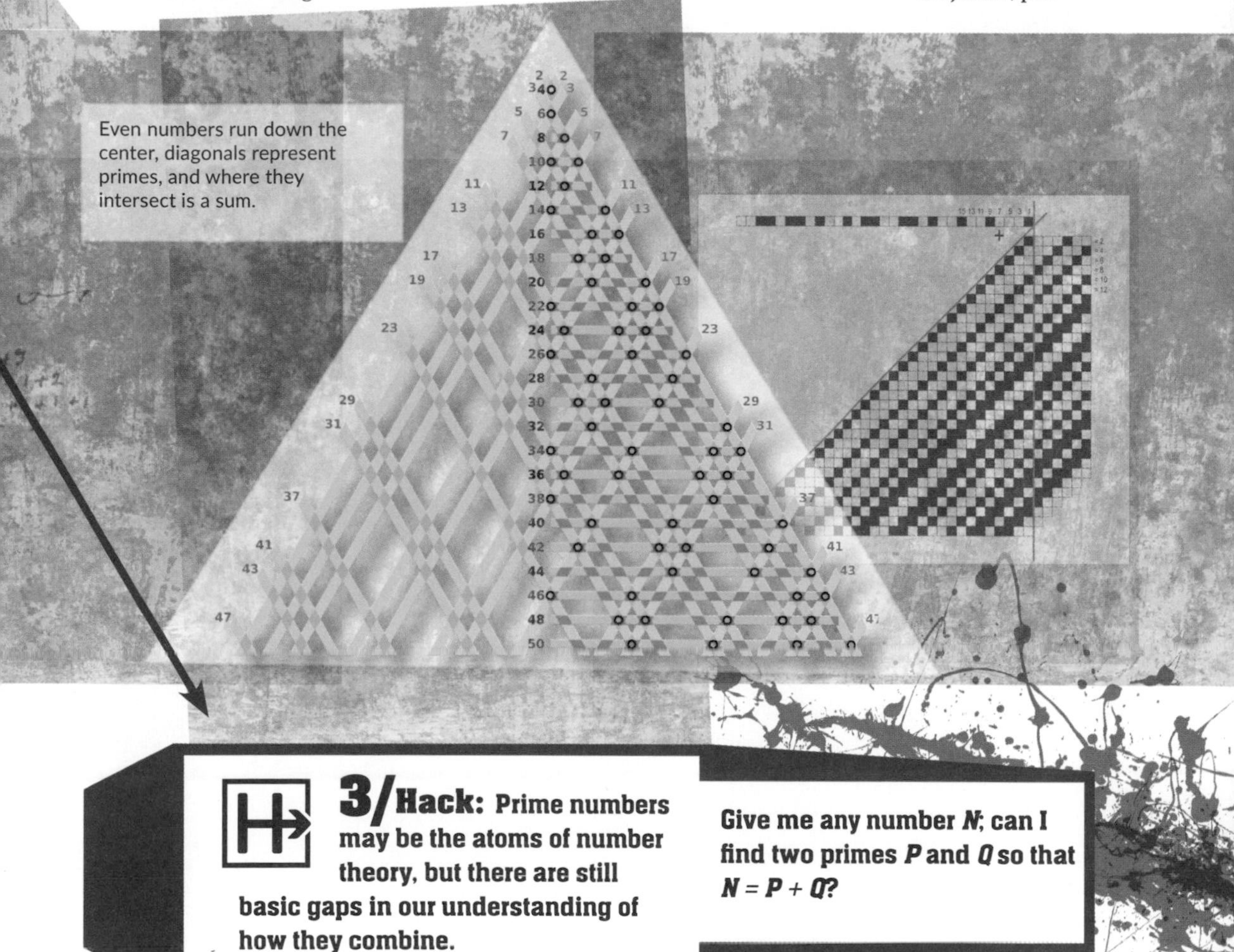

Even numbers run down the center, diagonals represent primes, and where they intersect is a sum.

3/Hack: Prime numbers may be the atoms of number theory, but there are still basic gaps in our understanding of how they combine.

Give me any number N; can I find two primes P and Q so that $N = P + Q$?

No.20 Negative Numbers

Less than zero

1/ Helicopter view: Suppose I have three apples and you try to take away five of them; most of us will agree this is impossible. Even today, when a young child is asked for the solution to 3 – 5, they may tell you it "cannot be done." They are absolutely right since this problem has no **natural number** solution.

To reply that 3 – 5 = –2, one needs to accept that numbers no longer represent quantities of objects. They have become more abstract and can represent, for example, credits and debits in finance or temperatures measured on an arbitrary scale.

The integers are what you get if you extend the natural numbers to include negative (whole) numbers and zero. They are a much more recent invention than the natural numbers, and less universal. Although they were used in India and China in antiquity, in the West the idea of a negative number was treated with suspicion as late as the time of Newton in the early 1700s.

Above: Ancient Indian and Chinese mathematicians independently discovered negative numbers, but they took much longer to be accepted in Europe.
Right: An ancient Chinese writing system that can represent positive and negative whole numbers and uses an empty space to represent zero.

2/ Shortcut:

2/Shortcut: Integers arise from the need to subtract a larger number from a smaller one and obtain a definite answer. Apart from 0, every integer has a size and an *orientation*—positive or negative—often called the number's *sign*.

Adding 0 to any number leaves it unchanged, and given any integer we can add the same integer with the opposite sign and get back to 0. These nice structural properties technically make the integers a special kind of set called a *group* under the operation of addition.

See also //
7 Set Theory, p.18
14 Natural Numbers, p.32
21 Rational Numbers, p.46
38 Groups, p.80

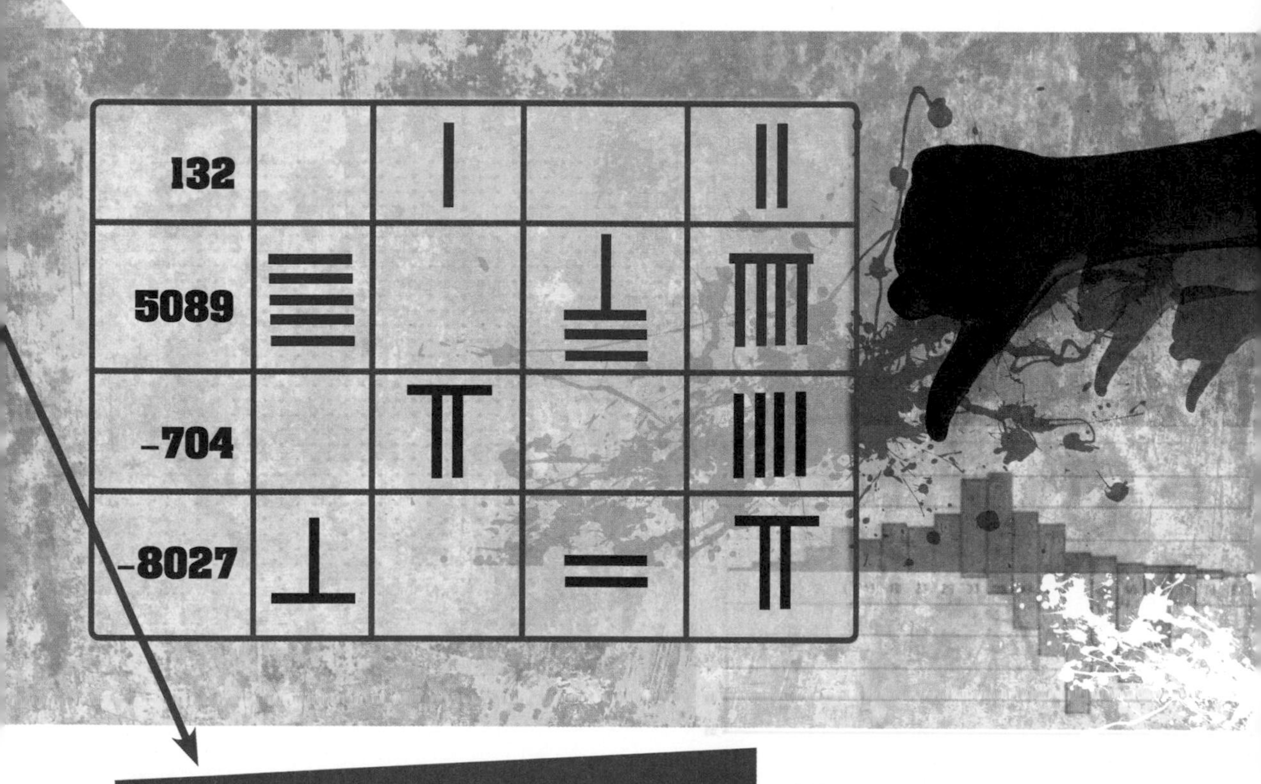

3/ Hack:

3/Hack: Negative numbers are natural numbers with an orientation, either positive or negative, along with 0, allowing us to do subtraction consistently.

Multiplying by –1 reverses a number's sign without changing its size.

No.21
Rational Numbers
Fractional improvements

1/ Helicopter view: The **natural numbers** are great for dealing with some things, especially addition and multiplication. If I want to divide 15 children into two equal groups, it makes sense to say this cannot be done: 15 ÷ 2 has no natural number solution. But can't I share 15 cakes between two people? I can give them 7 cakes each, leaving one over, which I can then cut in half. In such situations we need *rational numbers*.

We usually think of a rational number as a *ratio* of two integers, written as a fraction with one integer on top of the other. But, in fact, a single rational number can be represented as a fraction in various ways. For example, we can write $\frac{2}{3}$ as $\frac{4}{6}$, $\frac{6}{9}$, and in infinitely many other ways; these are all equivalent notations for the same object. The number 0 is represented by any ratio with 0 on top, such as $\frac{0}{1}$, $\frac{0}{2}$, $\frac{0}{3}$, and so on.

Above: Fractions are ratios of whole numbers; sometimes the square root of a fraction is a fraction, other times not.
Right: Dividing things up into equal parts is a basic operation in many areas of life, including music.

2/ Shortcut:

Rational numbers inherit their addition and multiplication properties fairly straightforwardly from the integers. What they add is the ability to carry out any division apart from one: division by 0.

This bothers some people enough to make them try to solve the problem, but it cannot be solved, at least not without surrendering a lot of basic algebra. The number 0 is important for addition and subtraction to work properly, but anything with its properties is bound to behave badly if you try to divide by it.

See also //
14 Natural Numbers, p.32
20 Negative Numbers, p.44
25 Irrational Numbers, p.54
38 Groups, p.80

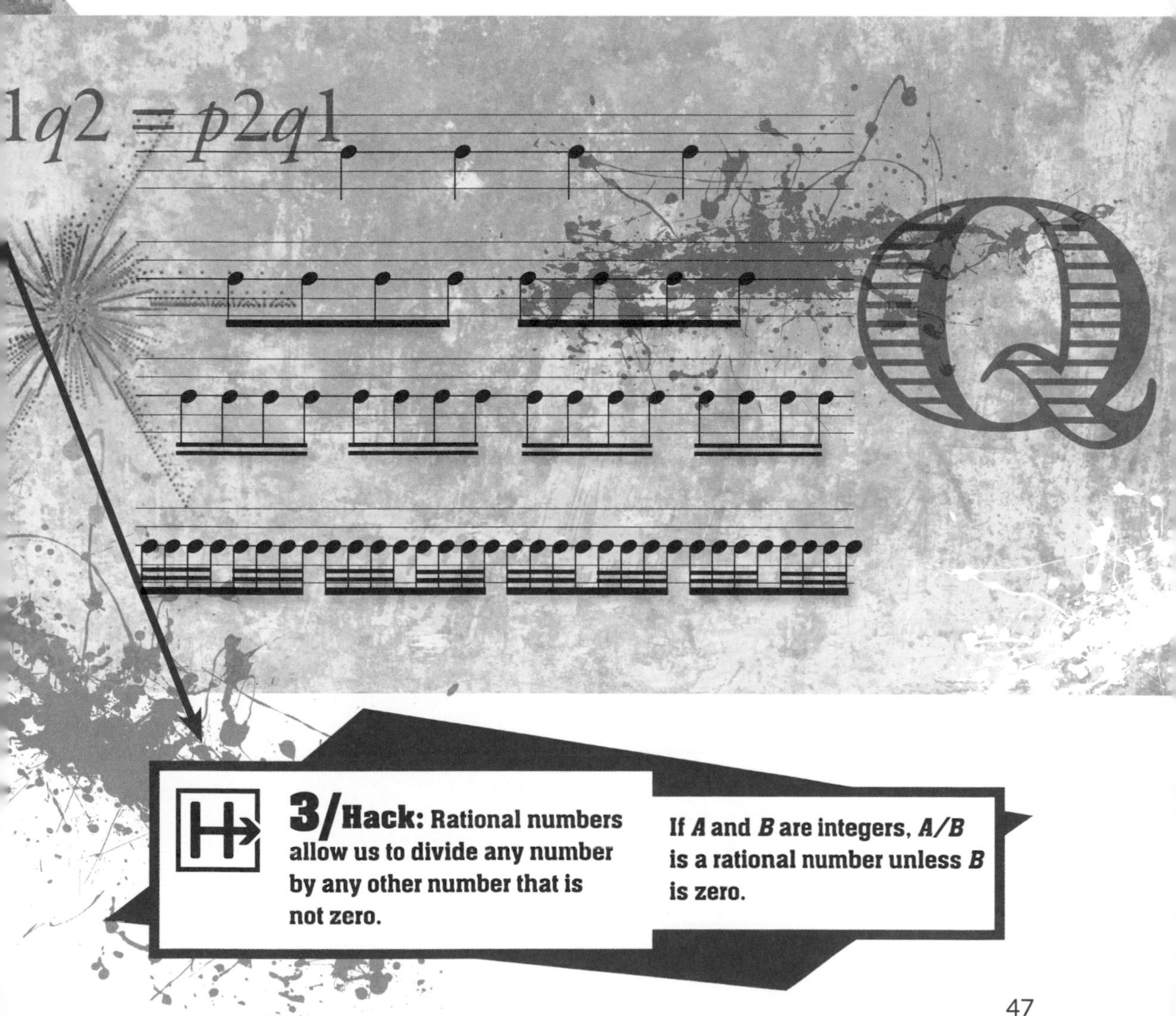

3/ Hack: Rational numbers allow us to divide any number by any other number that is not zero.

If *A* and *B* are integers, *A/B* is a rational number unless *B* is zero.

No.22 Powers Multiplication, multiplied

Mūsā al-Khwārizmī (*c.*780–850) made many important breakthroughs in algebra.

1/ Helicopter view: The area of a square is found by multiplying the length of one of its sides by itself; if L is the length, L^2 is the area. The volume of a cube is found by doing this for one face, then multiplying by itself *again*, which altogether is written L^3. This repeated multiplication-by-itself is the most basic version of the idea of a *power*, and is of ancient pedigree.

The modern notation took a very long time to develop, however, and this obscured the unity and breadth of the idea. It can be extended to negative and fractional powers, and even further into other number systems, yielding many useful and elegant results. Like **products**, the act of raising to a power (also called *exponentiation*) can even be generalized beyond numbers to apply to a wide range of mathematical objects.

Right: Natural number powers relate to area, volume and their higher-dimensional relatives.

2/ Shortcut:

2/Shortcut: Let A be any number. If n is a **natural number**, A^n is just A multiplied by itself n times. We can prove that $A^n \times A^m = A^{n+m}$, which leads us to define the special cases $A^1 = A$ and $A^0 = 1$. Those definitions are conventions that make what we just proved true for all n and m.

Now suppose $A^2 \times A^m = A^0$. For consistency, it must be that $m = -2$. Basic algebra tells us that A^m is really just $1/A^2$, so we think of negative powers as "one over the positive power."

Here's another one for natural number powers: $(A^n)^m = A^{n \times m}$. Now suppose we have $(A^2)^m = A^1$. This happens when we square A and then take its square root, but $2 \times m = 1$ implies $m = ½$. So we define fractional powers to be roots.

See also //
8 Products, p.20
14 Natural Numbers, p.32
20 Negative Numbers, p.44
21 Rational Numbers, p.46
25 Irrational Numbers, p.54
30 Transcendental Numbers, p.64

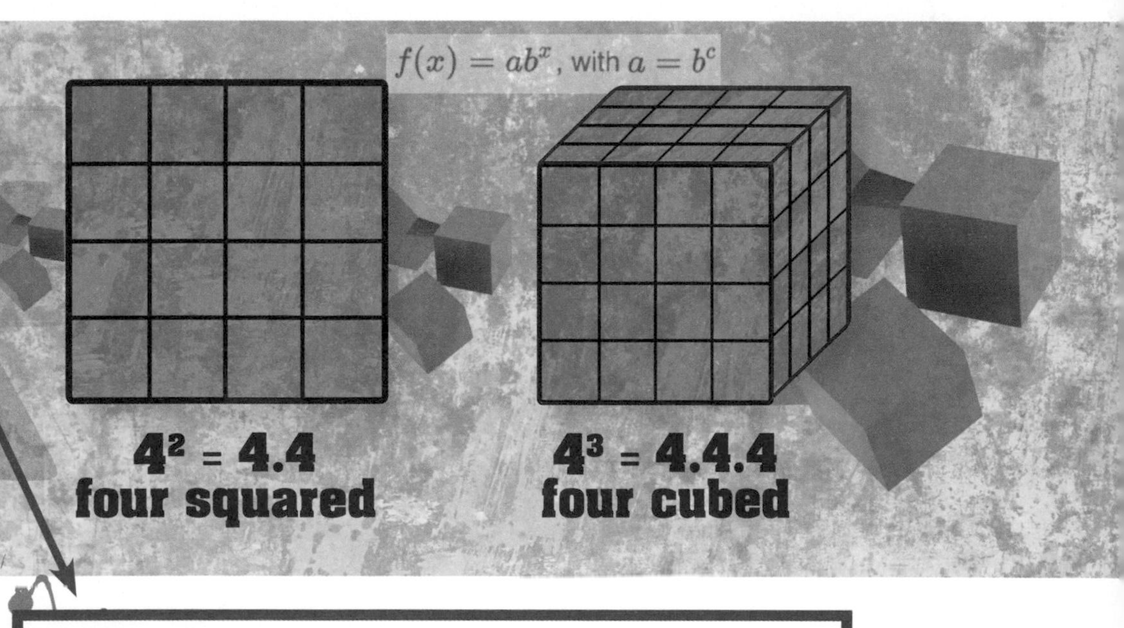

3/Hack: Raising a number to a natural number power is repeated multiplication; other types of power follow from the basic identities.

Xn: multiply X by itself n times
X-n: 1/Xn
X1/n: take the nth root of X

No.23 Polynomials Unknown quantities

1/ Helicopter view: *Polynomials* are fragments of algebra that look something like this:

$$c_0 x^0 + c_1 x^1 + c_2 x^2 + c_3 x^3 + \ldots$$

where the dots indicate the same pattern repeated forever.

The symbols c_0, c_1, and so on represent numbers that define the polynomial; these are called its *coefficients*. The symbol x is an "empty slot" that can be filled with any value we like, so it is called a *variable*. A coefficient times a power of x is called a term. We usually stipulate that all but a finite number of terms must be zero.

We can draw a graph of a polynomial by plotting how its value changes as the value of x changes. We can also set the polynomial equal to a number, forming an equation, and try to find a value of x that makes the equation true, which is called *solving* the equation.

Many mathematical expressions turn out to be polynomials; others are more complex but can be accurately approximated by them.

Right: The Sydney Harbor Bridge uses a polynomial curve (a parabola) for its lower arch.

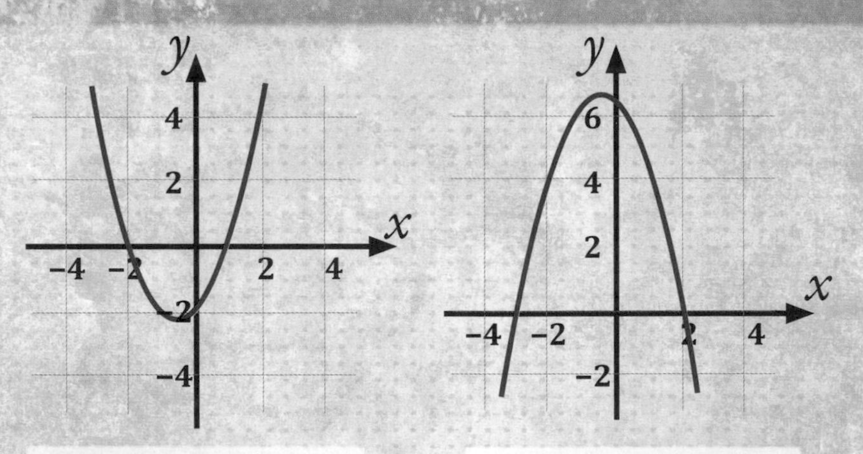

Even degree polynomial with a positive leading coefficient

Even degree polynomial with a negative leading coefficient

2/ Shortcut: Here is a recipe to make a *polynomial* in one variable with integer coefficients. Start with the symbol x to represent the variable.

Now consider all the natural number powers of this symbol: x^0, x^1, x^2, x^3, and so on, going on forever. Multiply each by an integer of your choice. Add all the terms together to make the final polynomial.

Note that we could use something other than integers to multiply our powers of x, and we can use more than one variable (x and y, say) to get more complicated examples.

See also //

9 Maps, p.22
14 Natural Numbers, p.32
22 Powers, p.48
25 Irrational Numbers, p.54
30 Transcendental Numbers, p.64
87 Varieties, p.178

3/Hack: Polynomials are maps from numbers (values of x) to other numbers made of only the simplest algebra; they appear very frequently in math.

Many phenomena are—or are similar to—polynomials, making their simplicity valuable.

No.24 Logarithms Powering down

1/ Helicopter view: Imagine I have a dish of bacteria, and the number of bacteria doubles every minute. I started with 1 of them. How long will it be before I have more than 100? Written as algebra, I want to find t such that $2^t > 100$.

Such problems are hard to solve without a new technique, known as the *logarithm*, invented in the early 1600s independently by John Napier and Joost Bürgi. To solve the problem just mentioned, take the base 2 logarithm of both sides. The result is approximately 6.644, which means we need to wait a little longer than 6 minutes 38 seconds for our bacteria to reach a population of 100.

But how do we calculate the number 6.644? That's not so easy. Initially it was done laboriously by hand, but today calculators do it in the blink of an eye. And it turns out that natural and mathematical processes alike often obey logarithmic laws of growth.

Above: Plotting the logarithm function in special coordinates yields a spiral that is sometimes found in natural structures.

2/Shortcut: Suppose we have an equation that looks like $A^n = B$. If we know A and n and want to find B, we use **powers** to do that. If we know n and B and want to find A, powers again will help us solve it.

But what if we know A and B and want to find n? This is the problem that *logarithms* are designed to solve. The logarithm to the base A, written $\log_A$, transforms every number X to the *power you have to raise A to to get X*. For example, since $2^3 = 8$, it must be that $\log_2(8) = 3$.

See also //

22 Powers, p.48

25 Irrational Numbers, p.54

45 Fermat's Last Theorem, p.94

John Napier (1550–1617) and pages from his *Mirifici Logarithmorum Canonis Discriptio* (1614), on the invention of logarithms.

3/Hack: Logarithms solve equations where the unknown quantity is the power another number is raised to.

If $A^n = B$, $\log_A(B) = n$.

No.25 Irrational Numbers

When fractions aren't enough

1/ Helicopter view: Suppose you have a square whose area is 2 units; how long must its sides be? There is no question that such a square can exist, at least in the mathematical sense, and the length of the side is straightforward. By definition it is $\sqrt{2}$, the *square root* of the area, which must be somewhere between 1 and 2.

Since the time of Pythagoras, though, we have known that this number cannot be written as a fraction. The proof is a great example of **reductio ad absurdum**, and it demonstrates that there is no **rational number** that squares to give exactly 2. We call $\sqrt{2}$ an example of an *irrational number*.

It was not until the 19th century that irrational numbers were understood. They came to be thought of as infinite sequences of rational numbers that get closer and closer to the value sought. That is, they were thought of as **limits**.

Above: Pi, the ratio of a circle's circumference to its diameter, is one of the most famous and puzzling irrational numbers. *Right:* The irrationality of the square root of 2 posed a serious conceptual challenge to the ancient Greeks.

2/Shortcut: **Rational numbers** seem to define all possible quantities. If I ask how tall someone is, you can answer with a fraction that is as precise as we need.

One way to do this is to give a length to a certain number of decimal places. This is a rational number in disguise. The more decimal places you add, the more precise the answer gets.

But to write an irrational number this way needs an *infinite* number of decimal places with no repeating pattern in their digits.

See also //

3 Reductio ad Absurdum, p.10
4 Limits, p.12
20 Negative Numbers, p.44
21 Rational Numbers, p.46
22 Powers, p.48
26 Real Numbers, p.56

3/Hack: Irrational numbers live side-by-side with the rational numbers; you can think of them as infinitely long decimals.

A quantity that cannot be written as a ratio of whole numbers is called "irrational."

No.26
Real Numbers Making a point

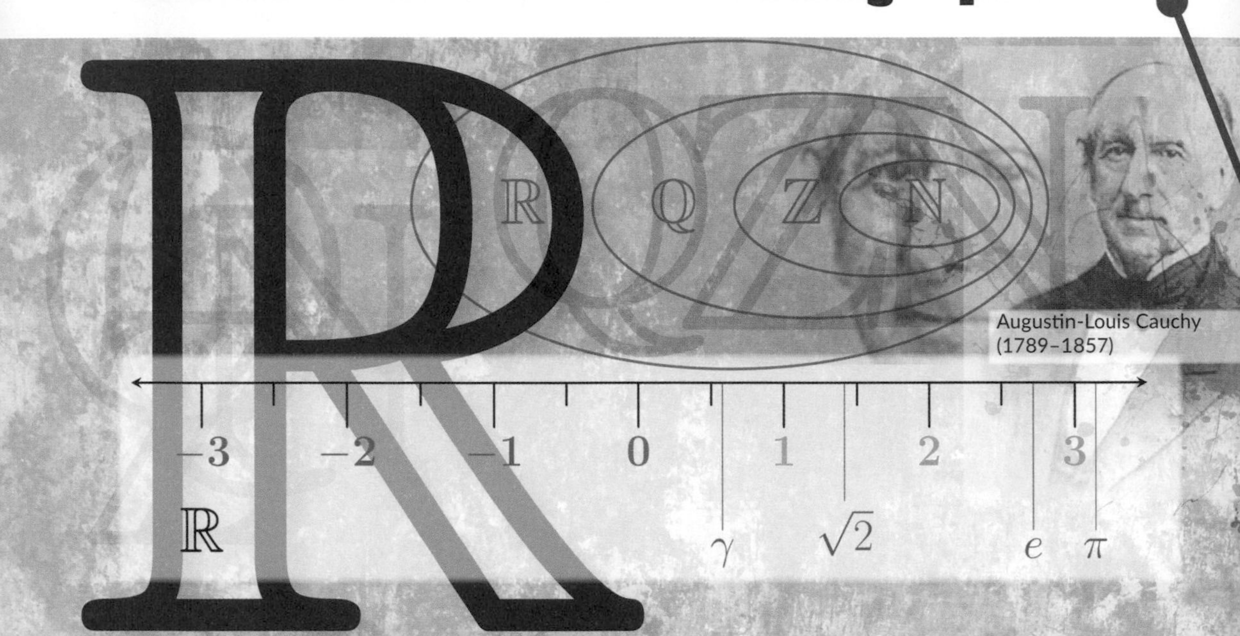

1/ Helicopter view: It looks as though the **rational** numbers could be used to label the points on a continuous, infinite line in the following way. Label an arbitrary point on the line 0, then decide on one direction as positive. Every point in that direction is labeled by the positive rational number indicating its distance from the 0 point, according to some unit of measurement. The **negative** rationals label distances in the opposite direction.

Unfortunately this doesn't cover all the points on the line. The ones that are missing are those whose distances from 0 are **irrational**. To turn all the points into numbers we need a number system that includes both rational and irrational numbers.

These are the so-called *real numbers*. They are used everywhere in mathematics and science because they provide a numerical representation of continuous space. They can do this because they have a property called *completeness*, meaning that every sequence of real numbers that could converge to a **limit** does converge, and the limit is again a real number.

Above: Real numbers (*R*) include the rational (*Q*), which include the integers (*Z*), which include the natural numbers (*N*).

2/Shortcut: *Real numbers* are defined in the same way **irrational numbers** are. Each is the limit of an infinite sequence of **rational numbers** that are said to approximate it. If the sequence actually arrives at its **limit**, the number is rational. If not, it is irrational.

It's possible to define ordinary arithmetic with real numbers, and even things like roots and **logarithms.** Whenever we start with a real number, we get an answer that is also a real number.

See also //
4 Limits, p.12
20 Negative Numbers, p.44
21 Rational Numbers, p.46
24 Logarithms, p.52
25 Irrational Numbers, p.54
33 Complex Numbers, p.70

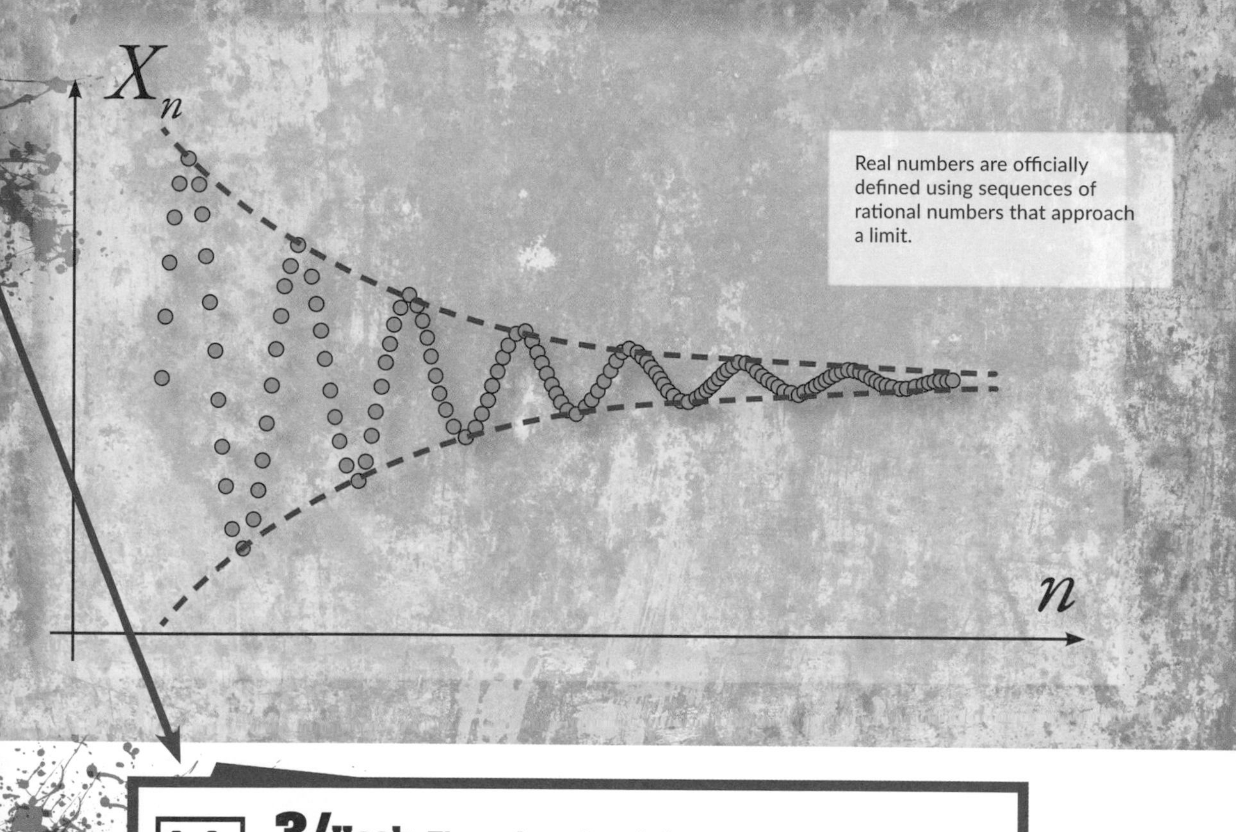

Real numbers are officially defined using sequences of rational numbers that approach a limit.

3/Hack: The real numbers bring together rational and irrational numbers to provide a distinct numerical label for every point on an infinitely long line.

A real number is the limit of an infinite sequence of rational numbers.

No.27

Cantor's Diagonalization Argument

Bigger and smaller infinities

Georg Cantor (1845–1914) was as shocked as his colleagues by his discovery of different sizes of infinity.

1/ Helicopter view: For centuries, infinity was a disreputable subject. Mathematicians blamed it for various lines of argument that had led to absurdities. It was not until Cantor's development of **set theory** that the idea began to be reclaimed.

Almost everyone accepts one infinite set, the set of all **natural numbers**, since so many mathematical ideas are impossible to formulate without it. Similarly, most accept endless sequences of numbers, where each number in the sequence can be identified by a natural number that indicates its position.

Now, consider infinite sequences of the digits 0 and 1. Suppose someone claims to have arranged all the possibilities in a sequential list, with each item identified by a natural number. Cantor shows this cannot be done; however constituted, the list must have missing sequences.

This implies the set of infinite sequences of 1s and 0s is *bigger* than the set of natural numbers. It is another, larger infinity! This applies to more than just 1s and 0s because the real numbers themselves can be subjected to the same procedure.

Right: Any method of listing real numbers (here depicted as binary sequences) misses out the number constructed from its diagonal.

2/Shortcut: The trick is to take the list—whatever it is and however produced—and show how to write down a sequence S that's missing from it.

The first digit of S is 0 if the first digit of the first entry in the list is 1; otherwise it is 0. The second digit of S is, likewise, the opposite of the second digit of the second sequence on the list.

Continuing in the same way produces a sequence that cannot possibly be on the alleged list.

See also //
3 Reductio ad Absurdum, p.10
7 Set Theory, p.18
12 The Schröder–Bernstein Theorem, p.28
14 Natural Numbers, p.32
26 Real Numbers, p.56
28 Infinite Cardinals, p.60

$$
\begin{aligned}
S_1 &= (1,\ 1,\ 1,\ 1,\ 1,\ 1,\ 1\ \ldots)\\
S_2 &= (0,\ 0,\ 0,\ 0,\ 0,\ 0,\ 0,\ \ldots)\\
S_3 &= (1,\ 0,\ 1,\ 0,\ 1,\ 0,\ 1,\ \ldots)\\
S_4 &= (0,\ 1,\ 0,\ 1,\ 0,\ 1,\ 0,\ \ldots)\\
S_5 &= (1,\ 1,\ 0,\ 0,\ 1,\ 1,\ 0,\ \ldots)\\
S_6 &= (0,\ 0,\ 1,\ 1,\ 0,\ 0,\ 1,\ \ldots)\\
S_7 &= (1,\ 1,\ 1,\ 0,\ 0,\ 0,\ 1,\ \ldots)\\
&\ldots\\
S_0 &= (0,\ 1,\ 0,\ 0,\ 0,\ 1,\ 0,\ \ldots)
\end{aligned}
$$

3/Hack: Cantor made it legitimate for the first time to speak of infinities, in the plural, and to compare their sizes.

Cantor's argument also implies there are strictly more real numbers than natural numbers.

No.28 Infinite Cardinals

Infinities that go on forever

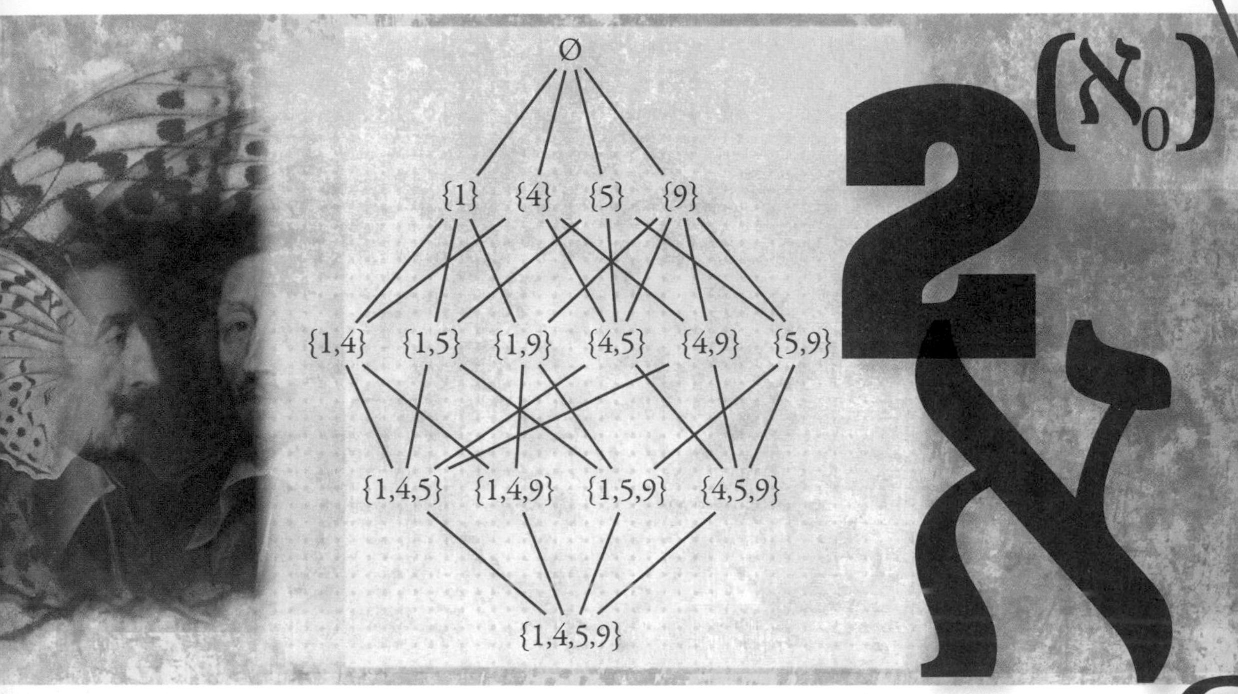

1/ Helicopter view: Cantor's diagonalization argument gives us two infinities, one larger than the other. It is natural to ask whether there are more, and how they are organized. In fact, Cantor found a way to produce as many different-sized infinities as we wish.

If S is a set, we can think about all the subsets that can be made from it. The collection of all these subsets is called the *power set* of S, written 2^S. Think of S as the items on sale in a shop and 2^S as the set of all possible shopping baskets you might have at the end of a visit.

The power set, 2^S, is always strictly larger than S. This can be seen directly when S is finite. A generalized version of the diagonalization argument shows it is also true for infinite sets; this allows us to produce ever bigger infinities simply by repeatedly forming the power set of the set we just created.

Above: The power set of a set always contains more elements than the original, even if it was infinite.
Right: The power set of the first five letters of the alphabet collects all the subsets you can make from them.

2/ Shortcut: Cantor wrote the size of the infinite set of **natural numbers** as $\aleph_0$. Now consider all the subsets of natural numbers: all the even ones; all the **prime** ones; all the ones greater than 7; and so on. The collection of all these is $2^{(\aleph_0)}$, which we write as $\aleph_1$.

The power set is always bigger than the original, in the sense of the **Schröder–Bernstein Theorem**. This means that if $\aleph_a$ is any infinite set, $2^{(\aleph_a)}$ is strictly larger.

See also //

7 Set theory, p.18

12 The Schröder–Bernstein Theorem, p.28

14 Natural Numbers, p.32

17 Prime Numbers, p.38

27 Cantor's Diagonalization Argument, p.58

29 The Continuum Hypothesis, p.62

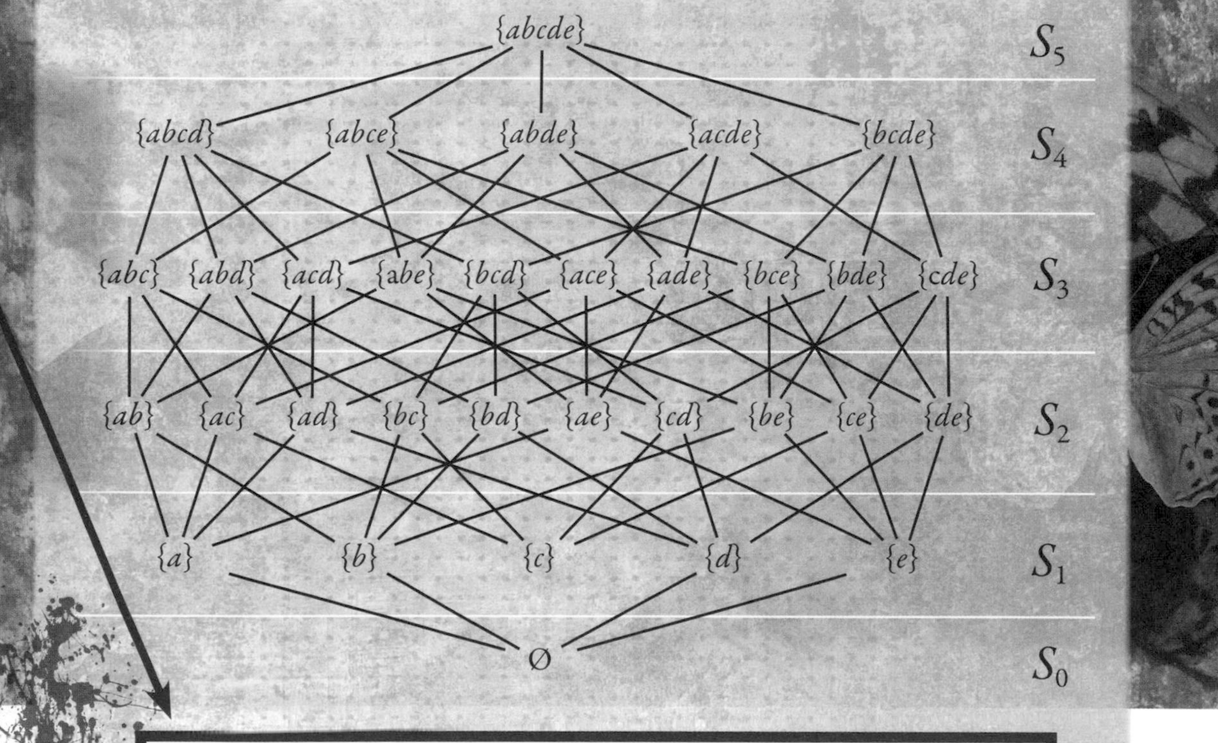

3/ Hack: The power set of a set is the collection of all its subsets. It always contains more elements than the original set.

Starting with an infinite set and repeatedly constructing power sets produces an unending sequence of ever larger infinities, known as the *infinite cardinals*.

No.29 The Continuum Hypothesis

How many points are on a line?

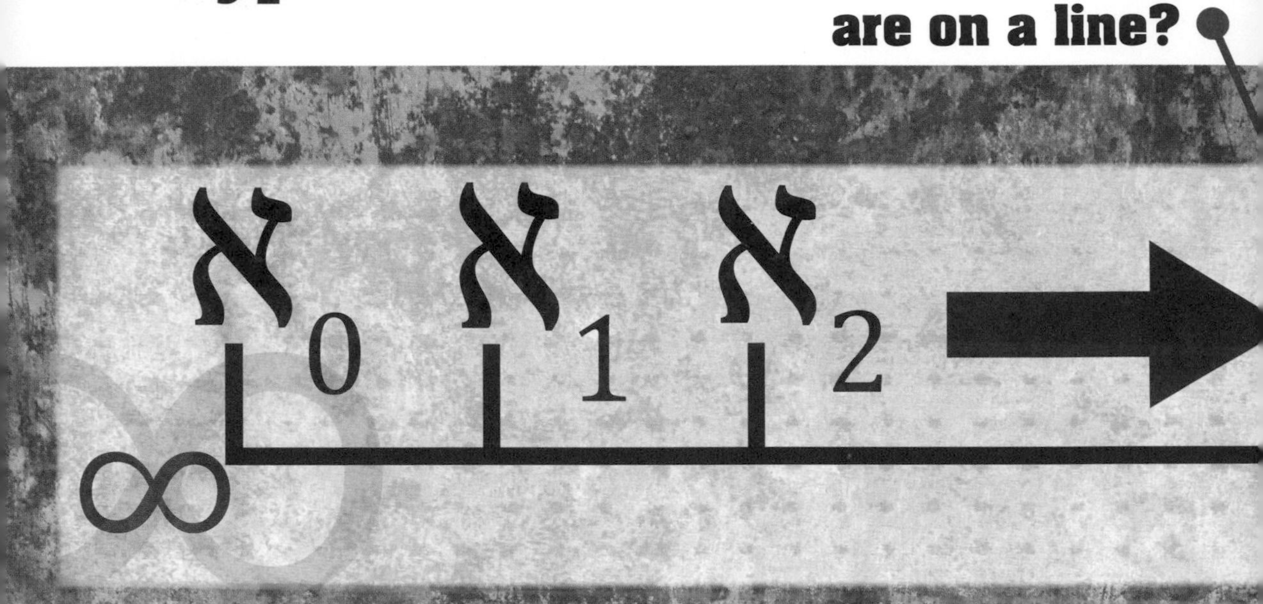

1/ Helicopter view: The real numbers can be used to label every point on a continuous line, leaving no gaps. Points seem much more mysterious than numbers, so this promises to give us some insights.

One question we can ask is, "How many points are on a line?" The number must of course be infinite. By **Cantor's diagonalization argument**, we know it must be strictly bigger than $\aleph_0$, the number of **natural numbers**.

The continuum hypothesis suggests that this number is the very next infinity up, $\aleph_1$, so that there is no size of infinity in between $\aleph_0$ and the size of the real numbers.

In 1940 Kurt Gödel showed that the *continuum hypothesis* does not contradict anything in the foundations of **set theory**, so it *could* be true. But in 1963 Paul Cohen proved that if the continuum hypothesis is false, that also creates no contradictions.

Technically, this means it is independent of set theory; it cannot be decided either way.

Cantor discovered not one or two infinities, but an infinite succession of them, each bigger than the last.

2/Shortcut: We know that there is no **natural number** between 0 and 1; they are immediate neighbors. Similarly, we define $\aleph_1$ to be the very next size of infinity after $\aleph_0$, and so on for $\aleph_2$, $\aleph_3$, and the rest.

The undecidability of the *continuum hypothesis* puts a limit on our ability to map this ladder of infinities onto more concrete mathematical objects. It has led some to conclude that these concepts may lie beyond the grasp of mathematics, at least as it stands today.

See also //
7 Set Theory, p.18
14 Natural Numbers, p.32
26 Real Numbers, p.56
27 Cantor's Diagonalization Argument, p.58
28 Infinite Cardinals, p.60

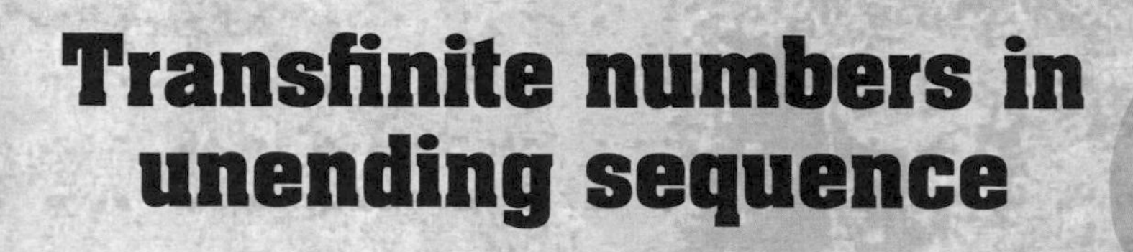

3/Hack: The continuum hypothesis takes mathematics to, and perhaps beyond, the edge of what it can do by posing a question and proving it cannot be answered.

It's impossible to identify which infinite cardinal represents the size of the real numbers.

No.30 Transcendental Numbers

What algebra can't do

1/ Helicopter view: Some **irrational** numbers can be described as solutions to equations involving **polynomials**. For example, if $x^2 - 2 = 0$ then we define $\sqrt{2}$ and $-\sqrt{2}$ to be the two possible solutions.

If x is the solution to any polynomial equation, it is called an *algebraic number.* The question is, are all real numbers algebraic? The answer is no, and those that are not algebraic are known as *transcendental numbers.*

The most famous transcendental number by far is π (*pi*), which is the ratio of a circle's circumference to its diameter. This was not proved to be transcendental until 1882, although it had been suspected for more than a century. Several other special classes of number have been shown to be transcendental, but not many, and in general the process is difficult.

The surprise is that Cantor showed there are as many **natural numbers** as there are algebraic numbers. Since there are strictly more real numbers altogether (by his diagonalization argument), in a sense "most" real numbers are transcendental.

Pi is transcendental, meaning it can't be the solution to any polynomial equation.

2/Shortcut: An algebraic number is the solution to the equation we make by setting a **polynomial** equal to zero. The integers are solutions to simple equations like $x - 8 = 0$. Here, of course, $x = 8$. Similarly, the **rationals** come from equations like $2x - 1 = 0$: here $x = \frac{1}{2}$.

Once **powers** of x appear in the equations, we get roots too, which are usually **irrational**, but polynomials *cannot* produce all irrational numbers in this way. The ones that cannot be solutions of polynomial equations are the *transcendental numbers.*

See also //

21 Rational Numbers, p.46
22 Powers, p.48
23 Polynomials, p.50
25 Irrational Numbers, p.54
26 Real Numbers, p.56
27 Cantor's Diagonalization Argument, p.58
28 Infinite Cardinals, p.60
31 Pi, p.66

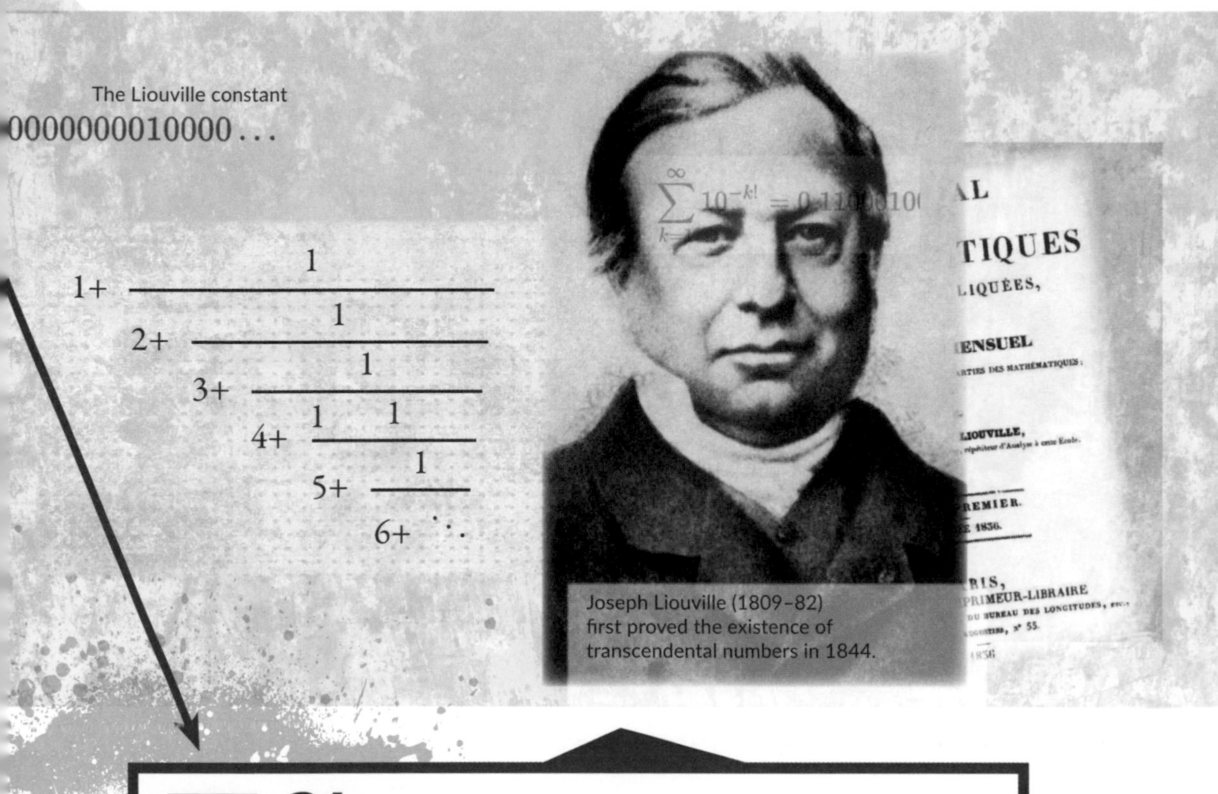

Joseph Liouville (1809–82) first proved the existence of transcendental numbers in 1844.

3/Hack: There is no easy, general way to represent the transcendental numbers which, in a sense, are the most common real numbers.

Algebraic numbers are solutions of polynomials. Transcendental numbers are not.

No.31

Pi The number that defines a circle

1/ Helicopter view: A circle's diameter is the distance from any point on it through its center to the point directly opposite. The circumference is the length of the circular line itself.

Suppose I have a rubber circle and make its circumference bigger by stretching it, keeping its circular shape. It is obvious that its diameter must also increase. In fact, every perfect circle has a special property; its circumference C and diameter D are related by the equation $C = \pi D$, where π (pronounced *pi*) is a number that is the same for every circle, no matter how big or small.

This fact has been known for thousands of years, and various attempts were made to calculate the value of *pi*. Good approximations were found, but no precise value. In fact, *pi* is a **transcendental number** and we do not know any way to describe it precisely that is not—pardon the pun—circular. Pure mathematicians have discovered a wealth of facts involving the number *pi*, many of which appear quite miraculous.

Above: Every circle's circumference is proportional to its diameter in exactly the same way, captured by the number *pi*.
Right: *Pi* has many strange mathematical properties.

2/Shortcut:

2/Shortcut: Circles are very basic shapes, and a great deal of mathematics is, in one way or another, geometry in disguise. So it's not surprising that π pops up repeatedly in almost every field, from physics to statistics.

We have developed several methods for approximating π for practical purposes. In 1995 Simon Plouffe devised a method for finding any given digit of π without calculating all the others before it, although in practice it is more like a computer program than a neat formula.

See also//
4 Limits, p.12
30 Transcendental Numbers, p.64
59 Euclidean Spaces, p.122

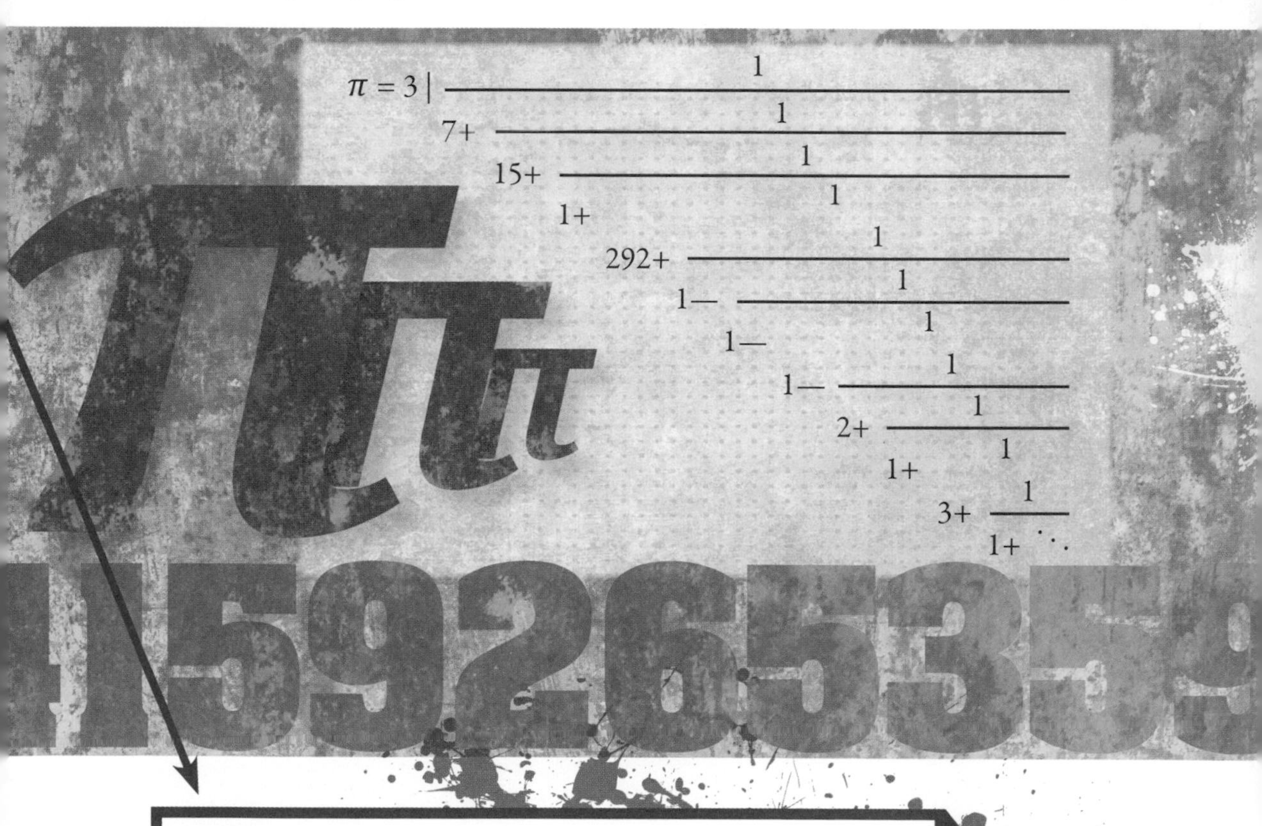

3/Hack:

3/Hack: Despite its simple origins, π contains deep mysteries and mathematical surprises.

The number π is the ratio of circumference to diameter, which all circles have in common.

No.32 Imaginary Numbers

What if mathematicians just made things up?

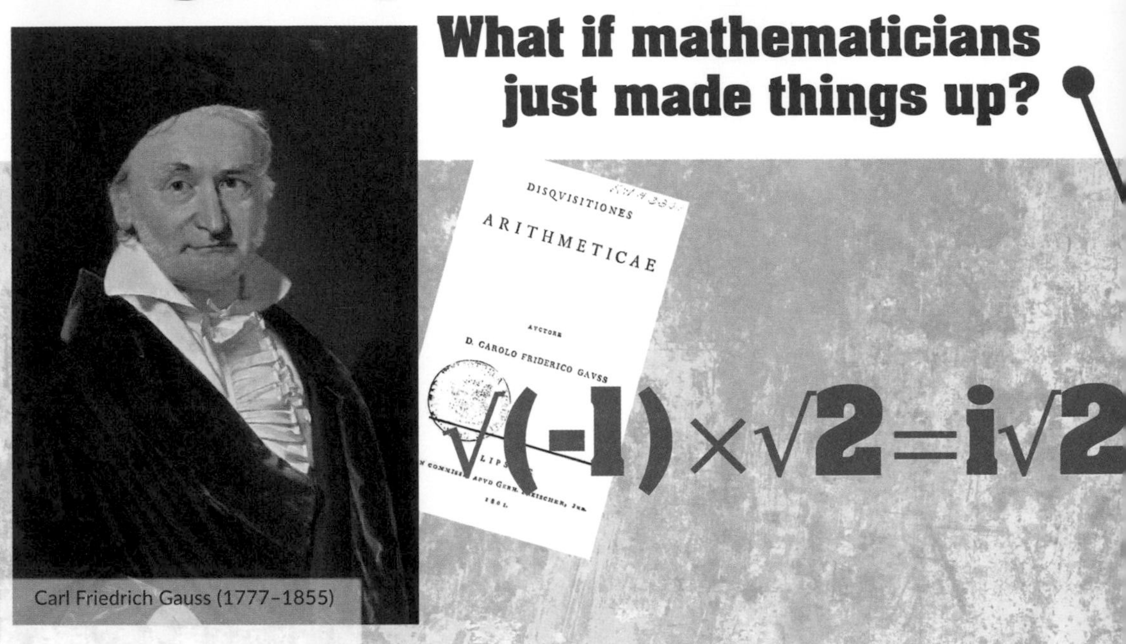

Carl Friedrich Gauss (1777–1855)

1/Helicopter view: Which numbers are **real**? That is a deep question. Many people feel there is something real about the **natural numbers**, but do **negative numbers** "exist" in the same way? What about **irrational numbers**? Each kind arises because the number system we are using does not contain solutions to a type of problem that we think ought to have a solution.

In the 16th century, Italian mathematicians developed numerous techniques for solving previously intractable algebra problems. For example, they knew they could define a number, $\sqrt{2}$, to solve the equation $x^2 = 2$. But what about the equation $x^2 = -2$? It seems as though $\sqrt{(-2)}$ should be a solution, but this number cannot exist.

To see why, remember the square root must be multiplied by itself to give the original number, so it should be that $\sqrt{(-2)} \times \sqrt{(-2)} = -2$. But $\sqrt{(-2)}$ must (it seems) be either positive or negative, and either way multiplying it by itself should yield a positive answer, not a negative one. Something new is needed.

Above: The use of imaginary numbers was not widely accepted until the work of Leonhard Euler and Carl Friedrich Gauss. *Right:* The graph of $y = x^3 - x$; imaginary numbers were used to find the places where it crosses the x-axis.

2/Shortcut: The new idea is to define a special symbol $i = \sqrt{(-1)}$. Then $\sqrt{(-2)}$ can be rewritten as $\sqrt{(-1 \times 2)}$, which the algebraic rules for roots tell us is $\sqrt{(-1)} \times \sqrt{2} = i\sqrt{2}$, a so-called *imaginary number*. But is this all just nonsensical symbol-shuffling?

The Italian algebraists found that sometimes they could use i in their working-out to avoid the blockage of the square root of a **negative number**, yet end up with a solution that does not mention i at all, and that can be independently checked. Niccolò Fontana Tartaglia (in the 16th century), for instance, discovered that 1, 0, and –1 are solutions of $x^3 - x = 0$ in just this way.

See also//
14 Natural Numbers, p.32
23 Polynomials, p.50
25 Irrational Numbers, p.54
26 Real Numbers, p.56
33 Complex Numbers, p.70

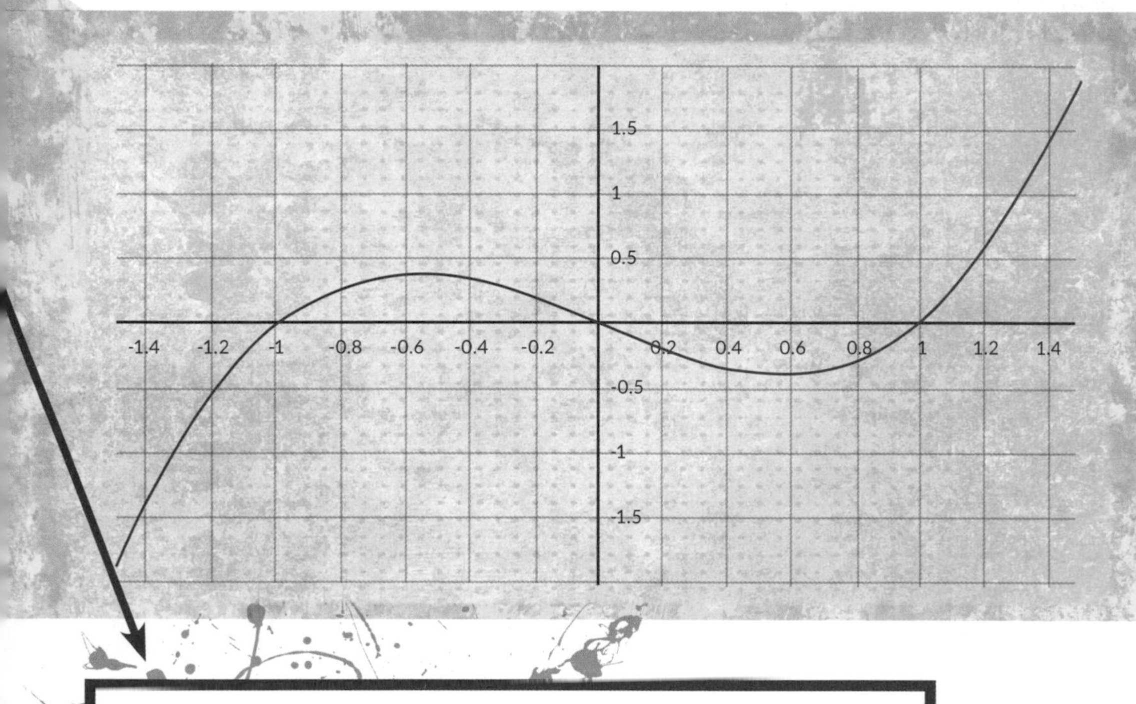

3/Hack: The square root of a negative number does not seem to make sense, because any ordinary number multiplied by itself gives a positive answer.

Defining i as $\sqrt{-1}$ helped Renaissance mathematicians overcome blockages in their algebra.

No.33 Complex Numbers

Arithmetic in two dimensions

1/ Helicopter view: The **imaginary numbers** floated around the margins of mathematical practice for two centuries before they were brought out into the open. They were a dirty secret, or a trick of the trade that sometimes worked but everyone suspected there was some cheating going on.

Once you have defined $i = \sqrt{(-1)}$, how might you treat it as a fully fledged number? We need to make it add, subtract, multiply, and divide with other numbers, respecting the usual arithmetic rules. The result is the set of *complex numbers*, which are used everywhere in mathematics, physics, engineering, and a variety of other fields.

Each is of the form $x + yi$, where x and y are **real numbers**. We can add and multiply them using standard algebra. We can also define what it means to apply things like powers, square roots, and logarithms to them. That requires a bit more work but the results are very convenient and well behaved.

Below: A complex number is a combination of a real part and an imaginary part, allowing it to be represented by a point in 2D space. *Right:* Many fractals are obtained by iterating maps involving the complex numbers.

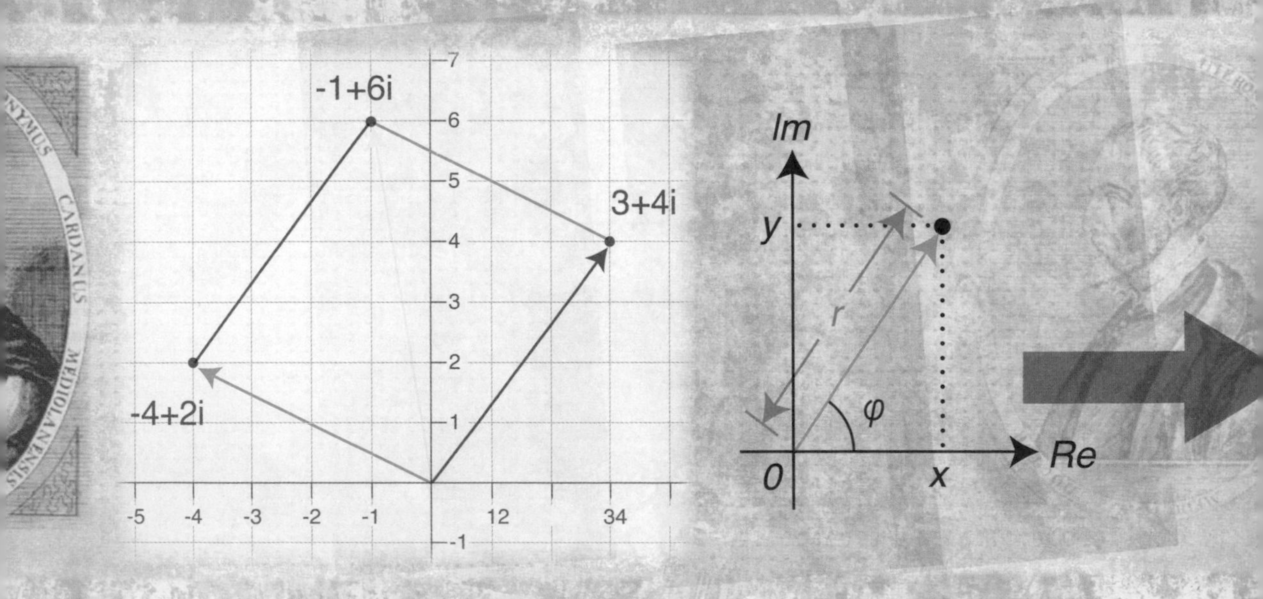

2/Shortcut:

The *complex numbers* are essentially two-dimensional. The number $x + yi$ has a real part, x, and an imaginary part, y, which are independent of each other. This means we can visualize a complex number as a point using a type of graph known as an Argand diagram.

An infinity of copies of the **real numbers** exists within the complex numbers, but complex numbers are much nicer to work with because they are algebraically closed. This means that every **polynomial** you can express with them can also be solved with them.

See also//
23 Polynomials, p.50
26 Real Numbers, p.56
32 Imaginary Numbers, p.68
42 Rings and Fields, p.88

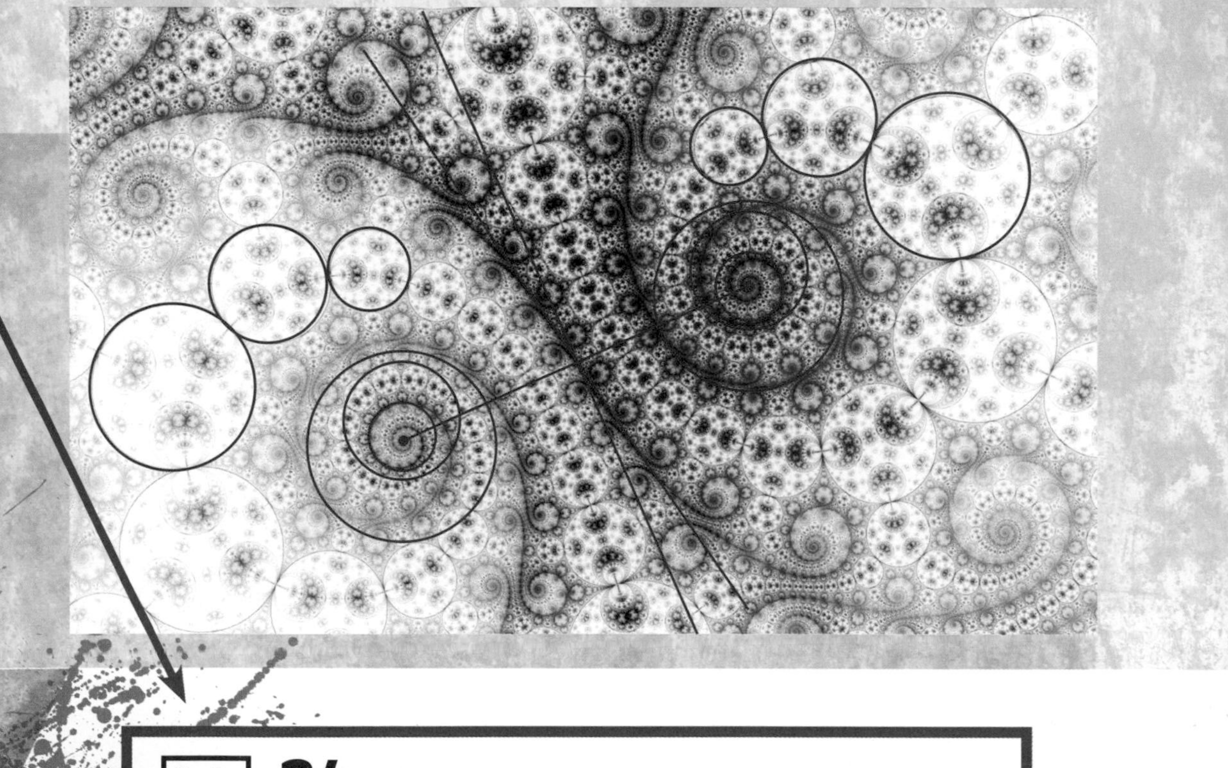

3/Hack:

The complex numbers allow us to solve problems that the real numbers do not; they are less intuitive but much neater.

Every complex number is $x + yi$; x and y can be any real numbers and $i = \sqrt{-1}$.

No.34

Quaternions

Are they numbers or rotations?

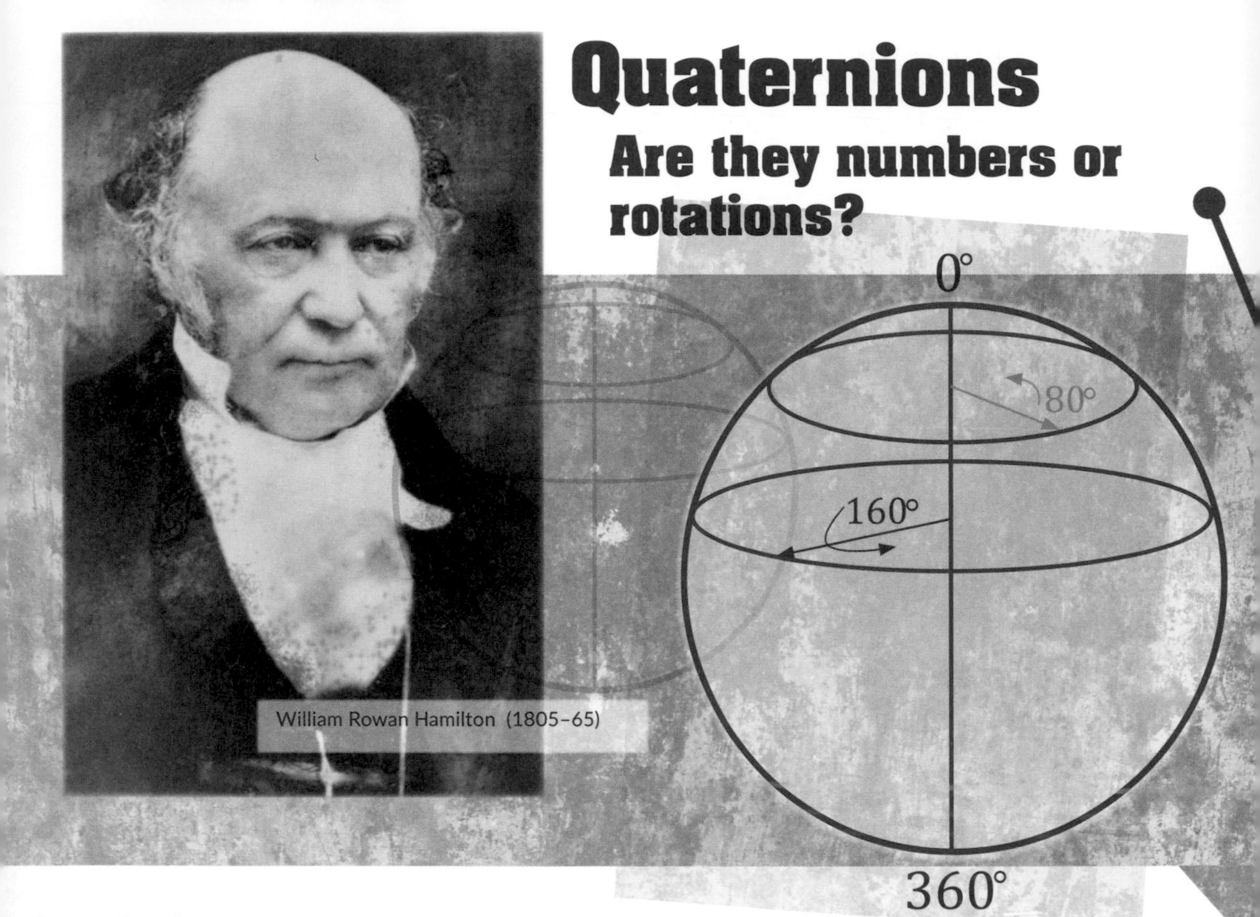

William Rowan Hamilton (1805–65)

1/ Helicopter view: In 1843, William Rowan Hamilton proposed an extension of the **complex numbers** that involved not one but three **imaginary** values: i, j, and k. All of these square to make −1, but they are not equal, and if you multiply them all together you also get −1. So every *quaternion* is of the form $w + xi + yj + zk$, where w, x, y, and z are **real numbers**.

Pure curiosity may have been the main motivation for Hamilton's bizarre-looking idea, but it turns out to have very elegant geometric properties. In particular, you can think of a quaternion as a way to rotate an object in 3D space, which has led to their widespread adoption in physics, robotics, aeronautics, and even computer games.

Sadly, these developments all came long after Hamilton's death. In his lifetime quaternions were never widely accepted.

Above: A visualization of rotations about a horizontal axis; long arrows rotate by bigger angles.

2/Shortcut:

You can work out many basic properties of *quaternions* from the definition, which stipulates how i, j, and k combine to produce –1. One curious property you may discover in this way is that their multiplication is not **commutative**, that is, the order you multiply in matters. For example, it follows from the definition that ij = k but ji = –k. This does not happen in any of the other number systems we have considered.

To "see" a quaternion $w + xi + yj + zk$ as a rotation, simply forget about w and look at $xi + yj + zk$ as a three-dimensional (3D) **vector** (x, y, z). The direction it points is the axis of rotation and its length is the angle.

See also//

26 Real Numbers, p.56

32 Imaginary Numbers, p.68

33 Complex Numbers, p.70

37 Associative, Commutative, Distributive, p.78

57 Vectors, p.118

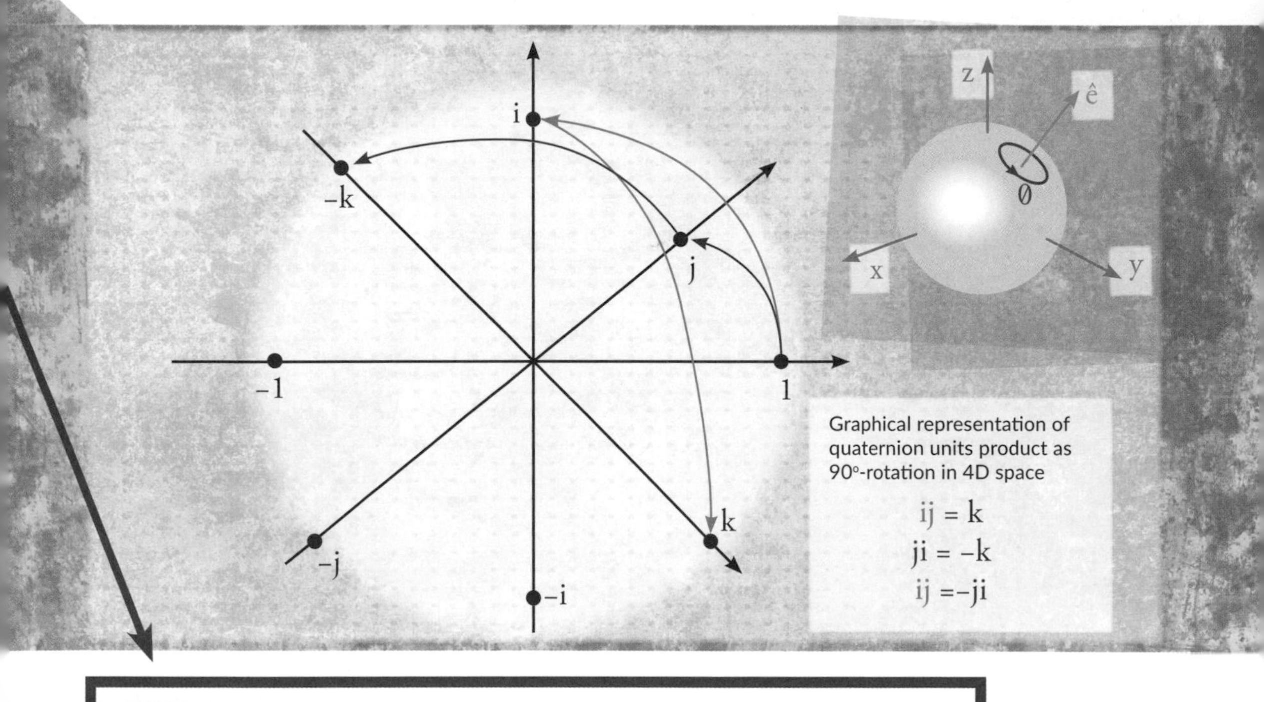

Graphical representation of quaternion units product as 90°-rotation in 4D space

$ij = k$
$ji = -k$
$ij = -ji$

3/Hack: Hamilton extended the complex numbers by adding two more imaginary square roots of –1; this turned out to be surprisingly useful in later applications.

The quaternions are a number system that can describe rotations in 3D space.

No.35 Abstract Algebra

It was never about numbers anyway

1/ Helicopter view: For most of its history, mathematics has been concerned with fairly concrete questions about numbers, shapes, or real-life practical problems. Many mathematicians work in this tradition today.

In the early 19th century, however, powerful new ideas emerged as a result of stripping away the content of these problems and looking, as far as possible, only at their formal or structural properties. These led to answers to questions that, in some cases, had gone unanswered for thousands of years.

This is the field we now call *abstract algebra*. Though it sounds fiendish, its abstraction actually makes it easier to learn. Gone are the things mathematics was supposed to be about, replaced by symbols that can mean anything we wish (a number, a shape, a process, a real-world object, a concept, and so on).

Because it is about pure structure, this approach has wide applications and can show us deep things we would not normally see.

Above: The permutations of Rubik's Cube form a group, a fundamental concept within abstract algebra.

2/ Shortcut: We begin with a **set** of objects (which could be anything) on which some extra "structure" is imposed. Usually this is done with **maps**.

For example, a **group** is a set with a special **binary operation** defined on it; the integers with the operation of addition have this structure. A **ring** (or **field**) has two binary operations and rules for their interaction. The structures of more complicated objects like vector spaces can be captured in similar ways.

See also//
7 Set Theory, p.18
9 Maps, p.22
36 Binary Operations, p.76
38 Groups, p.80
42 Rings and Fields, p.88
57 Vectors, p.118

The Cayley graph of the free group on two generators *a* and *b*.

3/ Hack: Using sets and maps, we can isolate and study structures independently of the things they are structuring, revealing hidden similarities between disparate subjects.

Abstract algebra distills the form of a theory by boiling away everything it seems to be about.

No.36
Binary Operations
Combining two into one

1/ Helicopter view: Perhaps the most fundamental structure in **abstract algebra** is the *binary operation*. It is something we define on a **set**, although what is in the set does not matter (that's what makes the algebra abstract). A binary operation takes two elements of the set, in order, and combines them to produce a third element from the same set.

Suppose your set contains every possible paint color; you could define a binary operation that takes two colors and produces another by mixing them in equal quantities. Or if your set consisted of the children in a school, you could define an operation to take any pair of children and give the elder child as the result.

Or consider addition. When we say $2 + 3 = 5$, we really mean the binary operation + is a **map** from the pair of elements $(2, 3)$ to 5. Exactly the same structure underlies subtraction, multiplication, and raising-to-the-power, as long as they make sense for all pairs of elements in the set.

A binary operation combines two elements of a set to produce another. Its action can be summarized by a "multiplication table."

2/Shortcut:

Technically, a *binary operation* on a set S is a **map** from the product $S \times S$ to S; that is, it maps every pair of elements of S to another element of S. There are no restrictions on how this should work, aside from the usual restrictions on maps in general.

Note that division is not a binary operation on any common number system. Dividing by zero is not allowed, so we cannot define a valid map from the Cartesian product $S \times S$ to S that represents division correctly.

See also//
7 Set Theory, p.18
9 Maps, p.22
35 Abstract Algebra, p.74
38 Groups, p.80

	f	g	h	k
f	f	f	f	f
g	f	g	h	k
h	k	h	g	f
k	k	k	k	k

3/Hack:

Arithmetic is made of binary operations, but so are other things in mathematics and everyday life; many mathematical structures are defined in terms of them.

A binary operation combines two things to produce another thing of the same type.

No.37 Associative, Commutative, Distributive Abstract properties

1/ Helicopter view: Suppose we have a set with a **binary operation** ¤. We write the elements of the set as a, b, c, and so on, and the binary operation as $a ¤ b$.

The operation is associative if $(a ¤ b) ¤ c = a ¤ (b ¤ c)$. This means it does not matter whether we figure out what $a ¤ b$ is first, then do the operation with c, or we figure out $b ¤ c$ first and then do the operation with a. Most important binary operations in mathematics are *associative*, but there are exceptions.

The operation is *commutative* if $a ¤ b = b ¤ a$. This means we get the same result regardless of the order we take a and b in. Addition of integers is commutative but subtraction is not.

Suppose we have another operation, written as §. If $a § (b ¤ c) = (a § b) ¤ (a § c)$ we say § *distributes over* ¤. In most number systems, multiplication distributes over addition, which leads to the method of "multiplying out brackets" that we learn at school.

Binary operations often share common properties that can be used to deduce information about them.

2/ Shortcut: As mathematicians turned away from objects toward the structures that objects live in (**abstract algebra**), they noticed some common properties recurring in fields of mathematics that had previously been thought quite distinct. Many of these, including *associativity*, *commutativity*, and *distributivity*, involve **binary operations**.

They are easy to state in symbols and often have consequences for what we can prove. Although special cases of them had been known for millennia, identifying and generalizing them has made new and sometimes profound patterns visible.

See also//
20 Negative Numbers, p.44
35 Abstract Algebra, p.74
36 Binary Operations, p.76
38 Groups, p.80
42 Rings and Fields, p.88

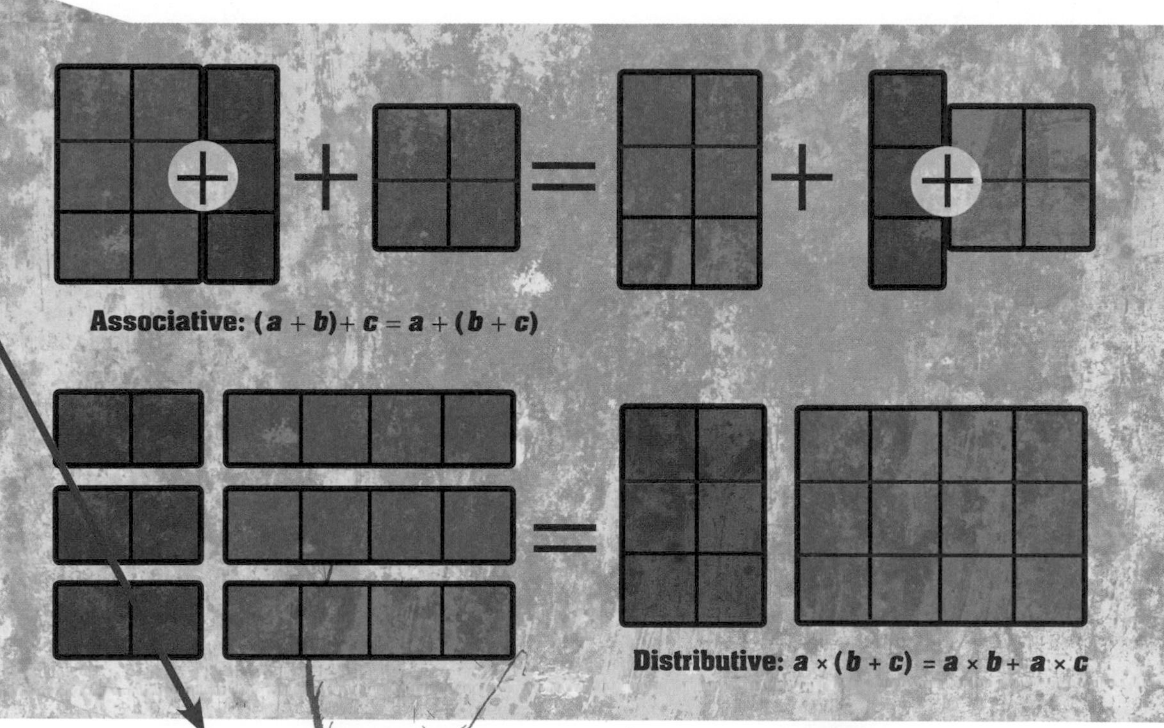

3/ Hack: A binary operation is:
- associative if it does not need brackets
- commutative if order does not matter
- distributive over another if we can multiply out brackets.

The properties of binary operations in abstract structures apply to many different areas of mathematics.

No.38
Groups The structure of symmetry

1/ Helicopter view: If a geometric shape can be transformed in some way so that it looks the same at the end of the process as it did before, we call the transformation a *symmetry*.

For example, an equilateral triangle can be rotated in two different ways that are symmetries, and there are three ways it can be flipped over (reflected) that leave it looking the same, too. Along with the transformation that does nothing—known as the *identity*—this makes a total of six symmetries.

A little experimentation shows that doing two of these in sequence produces the same effect as doing one of the others. Therefore, we can think of the set of symmetries as having a **binary operation** that combines any pair to produce another.

Investigating the properties of this operation leads us to define a *group*. The group structure turns up everywhere in mathematics and in many fields of science, too (see **representation theory** for some examples).

The symmetries of physical and geometric objects are among the most natural things to exhibit the group structure.

2/Shortcut: A *group* is a **set** with a **binary operation** that obeys a few structural rules based on the way symmetries behave.

Every symmetry can be reversed by a symmetry, which is called its **inverse**. For example, each rotation of the triangle undoes the other, and each reflection undoes itself.

Under this abstract definition, it turns out that addition of integers is a group, with each number's inverse being the same number with the opposite sign. That gives an indication of how widespread this structure is.

See also//
7 Set Theory, p.18
11 Inverses, p.26
35 Abstract Algebra, p.74
36 Binary Operations, p.76
42 Rings and Fields, p.88

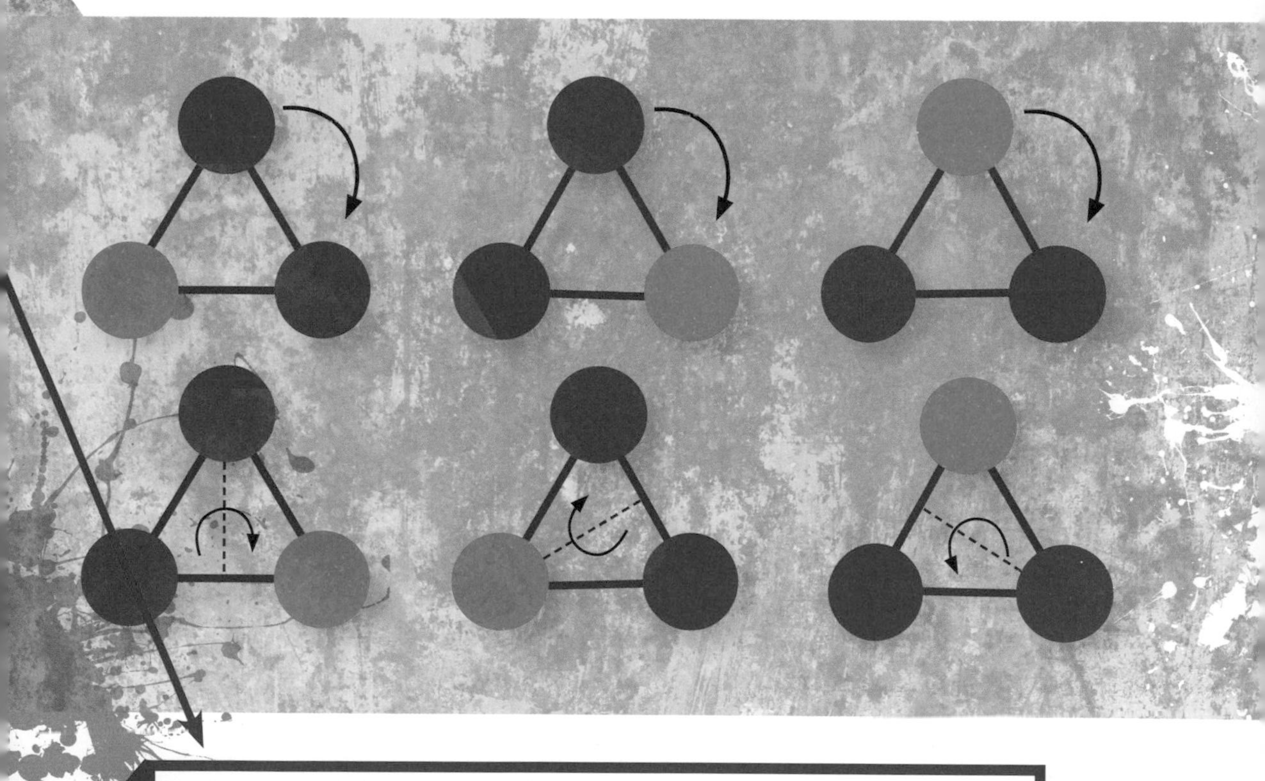

3/Hack: Groups generalize the notion of symmetry and allow us to apply it to objects that we had never previously considered to be "symmetrical."

Officially, a group is a set with an associative binary operation that has an identity and inverses.

No.39 Frieze and Wallpaper Patterns

Seventeen flavors of wallpaper

1/ Helicopter view: A visual pattern can be described by its **group** of symmetries (the rules that combine two symmetries to produce another). We sometimes even say two patterns with the same symmetry group are "essentially the same" (this is an **equivalence** relation).

Consider a flat plane that extends forever in all directions. Is there any limit to the symmetry groups that patterns on the plane can exhibit?

For a pattern that repeats forever in one direction only, like a *frieze*, there are just seven possible symmetry groups. For a pattern that repeats in both directions, like *wallpaper*, there are seventeen.

These extraordinary discoveries were made collaboratively in the second half of the 19th century. Similar results were also found in three dimensions, where they are called *crystallographic groups*.

In 1910 Ludwig Bieberbach proved that similar results can be found for spaces of all higher **dimensions**. In four-dimensional space there are 4,783 possibilities, a fact that was only determined in the 1970s. We do not know the numbers yet for higher dimensions.

Friezes are 2D patterns that repeat in one direction (e.g. horizontally), wallpapers repeat in both.

2/ Shortcut: What are the ways to move a pattern on a plane in an attempt to match it up with itself? We can slide the pattern around; this is called *translation*. We can also choose a point and rotate the pattern around that point. Finally, we can choose a line and reflect the pattern in it as if it were a mirror. One small complication aside (concerning glide reflections), those are all the possibilities.

These results—sometimes even the higher-**dimensional** ones—have a great many applications in science and technology because of the structural similarities they can reveal.

See also//
10 Equivalence, p.24
38 Groups, p.80
59 Euclidean Spaces, p.122
76 Dimensions, p.156

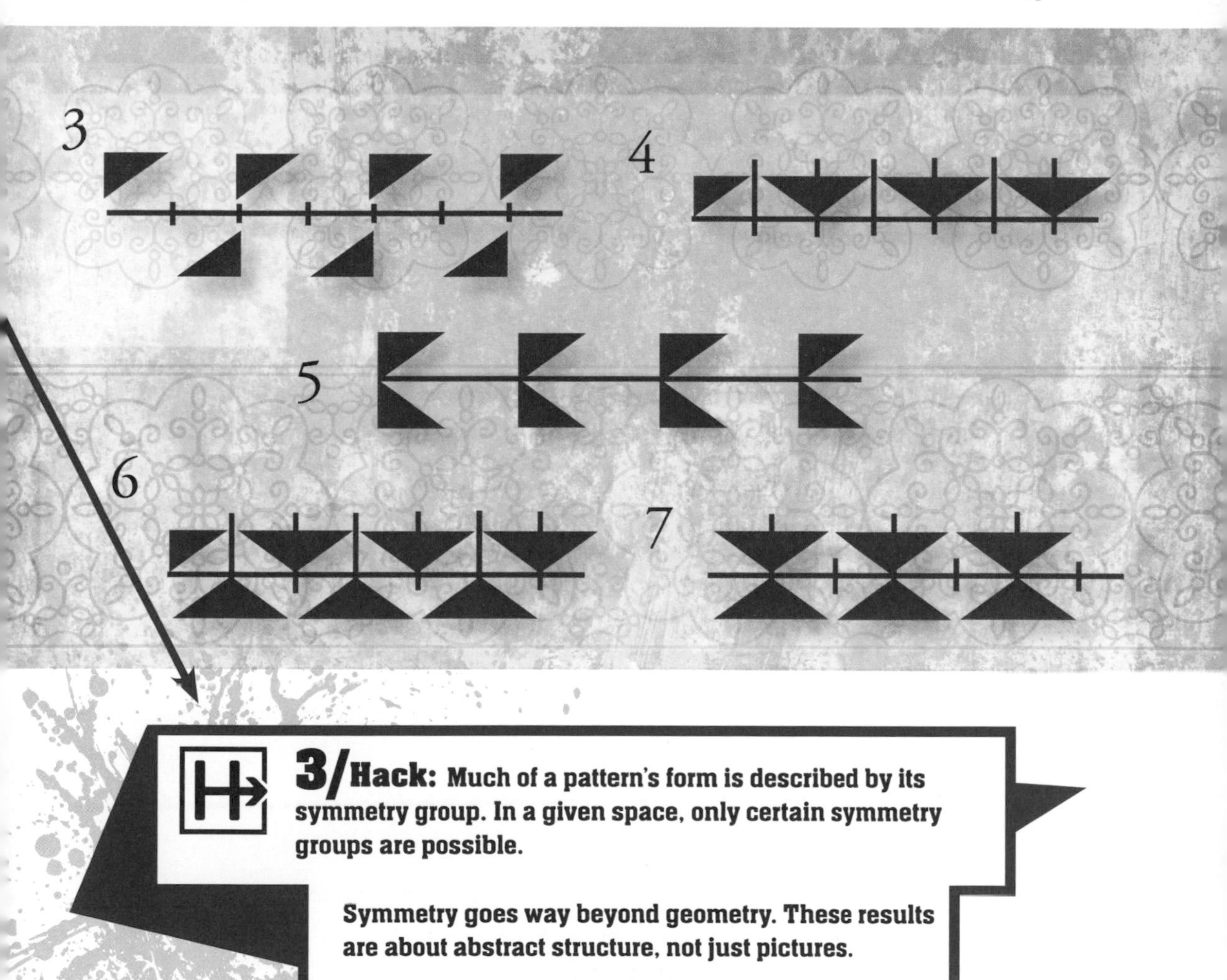

3/ Hack: Much of a pattern's form is described by its symmetry group. In a given space, only certain symmetry groups are possible.

Symmetry goes way beyond geometry. These results are about abstract structure, not just pictures.

No.40 Finite Simple Groups

The enormous theorem

1/ Helicopter view: Initially **groups** were used in a limited part of algebra, but soon mathematicians began to identify them everywhere. By the early 20th century it was clear that understanding their structure would yield insights across the subject.

If the **set** underlying a group is finite, we have a *finite group*. The group of symmetries of an equilateral triangle is finite (there are six of them) but the group of integers with the operation of addition is infinite.

Just as sets can have subsets, groups can have subgroups. Some have a special relationship with the bigger group. These are said to be *normal*. If a group G has a normal subgroup N, there is a procedure called *quotienting* that makes a new, smaller group called G/N. A group with no normal subgroups is called a *simple group*.

Simple groups cannot be "collapsed down" by quotienting, so they can be considered building blocks of group theory.

The cyclic groups containing a prime number of elements are examples of finite simple groups.

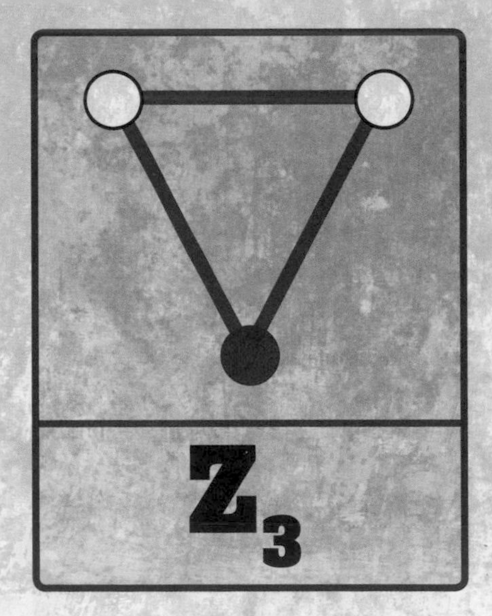

2/ Shortcut: We now know that the *finite simple groups* are a weird bunch. There are three main kinds, along with twenty-six so-called *sporadic groups*. These include the five *monstrous groups* which, though still finite, contain a huge number of elements.

This complete picture came out of a huge research project in the second half of the 20th century. Classifying the finite simple groups was a vast achievement that no individual mathematician could have accomplished alone.

See also//
7 Set Theory, p.18
38 Groups, p.80

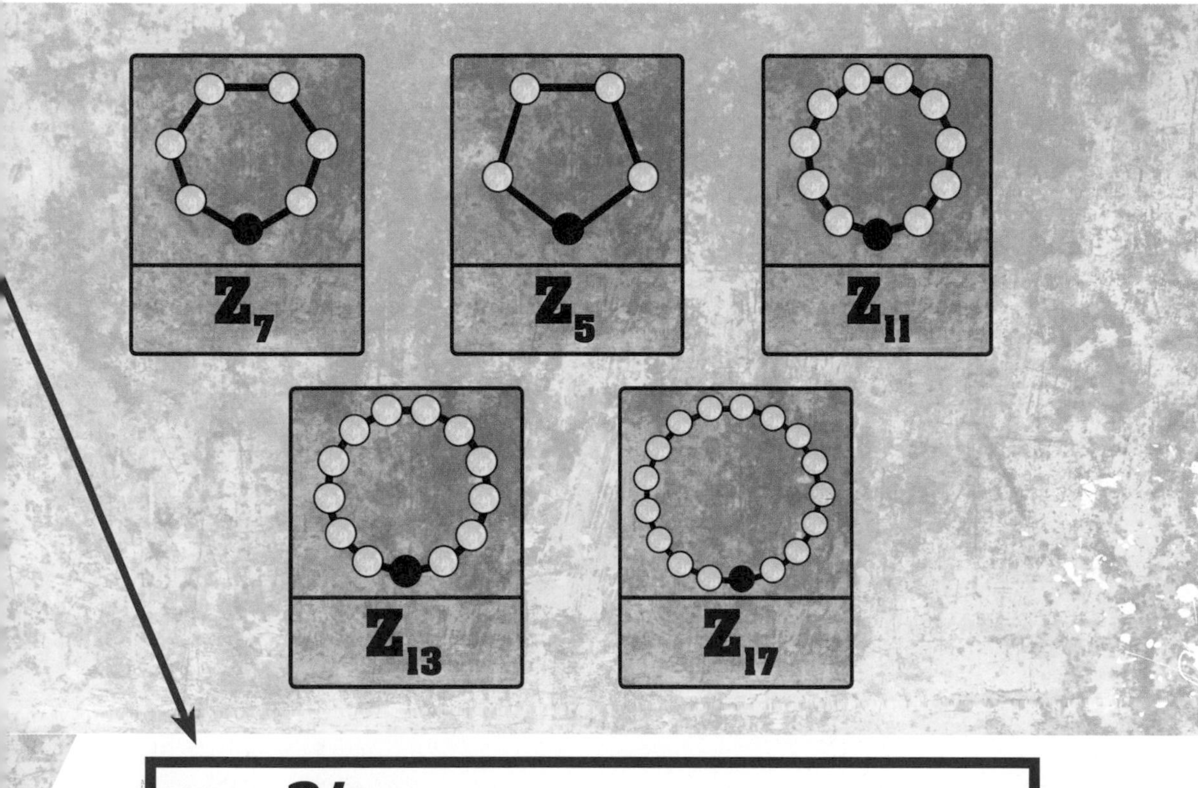

3/ Hack: Finite simple groups are like the prime numbers of group theory. We understand them quite well but they still have many secrets.

The classification of the finite simple groups is complete, but many questions about them remain open.

No.41

Lie Groups

Smooth symmetries

Lie groups are named after Norwegian mathematician Sophus Lie (1842–99).

1/ Helicopter view: **Groups** of symmetries can be finite, or at least involve discrete movements. The **frieze and wallpaper patterns** are governed by groups like these. They represent the kind of symmetry you learn about as a child, such as a few lines of reflection or points of rotational symmetry, for example.

What about a circle? Its symmetries look very different, because every line through its center is a line of symmetry, and every possible rotation about its center is a symmetry too. There must be an infinite number of them.

Looking just at the rotations, it is clear that we can make a rotation as small as we like, leading us to ask what happens in the **limit** when we reach something like an "infinitesimal rotation." The symmetries of a circle look more like smooth, flowing movements than sudden jumps.

Some groups do exist that capture such phenomena, and they are named *Lie groups* after mathematician Sophus Lie.

Above: Continuous symmetries are encountered frequently in physics.
Right: All the possible rotations of 3D space can be thought of as points on the (2D) surface of a sphere.

2/Shortcut: A *Lie group* represents the continuous transformations of a space—particularly, a **manifold**—that preserve some quantity or other. This preservation is what makes them symmetries.

The spaces studied in physics are usually manifolds and their laws can very often be stated in the form of conservation of some quantity (momentum, energy, charge, spin, and so on). A Lie group can be defined that shows everything that can happen in the space while conserving that quantity. Thus, a great deal of modern physics studies Lie groups.

See also//
4 Limits, p.12
38 Groups, p.80
39 Frieze and Wallpaper Patterns, p.82
60 Manifolds, p.124

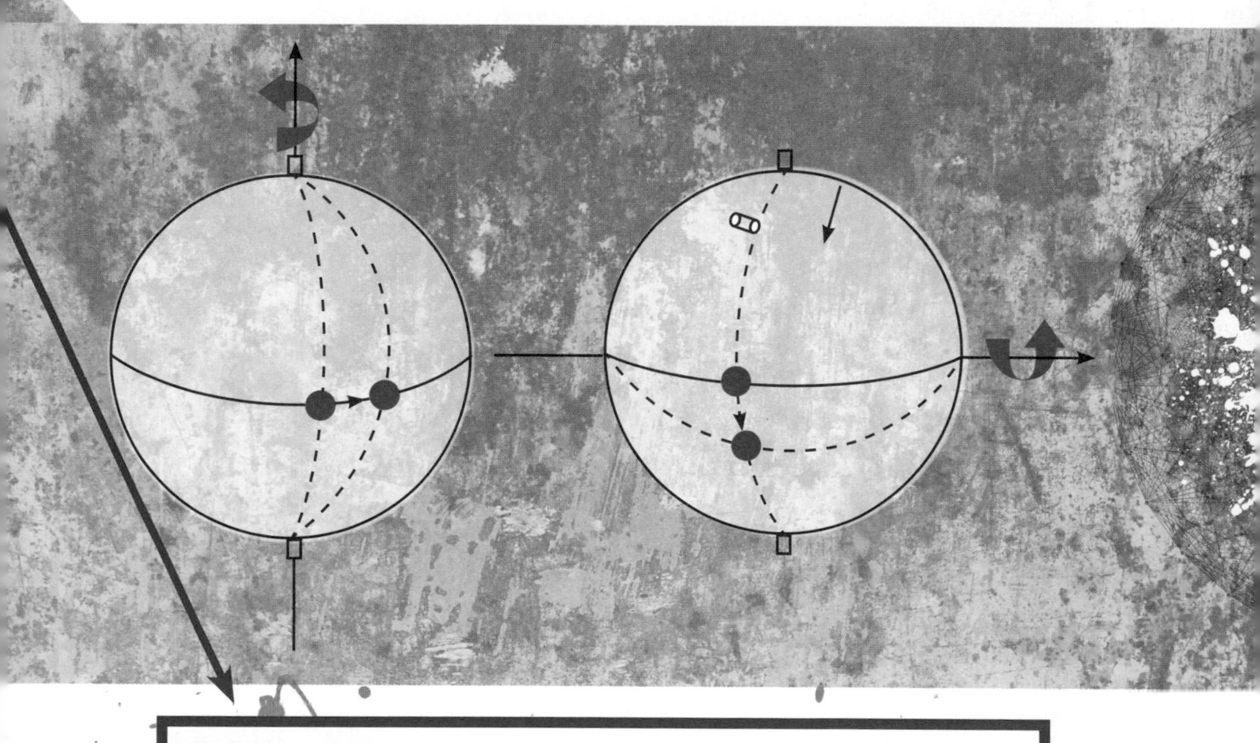

3/Hack: Lie groups capture continuous, smooth transformations of a space that preserve something; that is, in a sense, they are symmetries.

Lie groups are central to the study of manifolds, and to modern physics, and their theory is very well developed.

No.42
Rings and Fields
Arithmetic in general

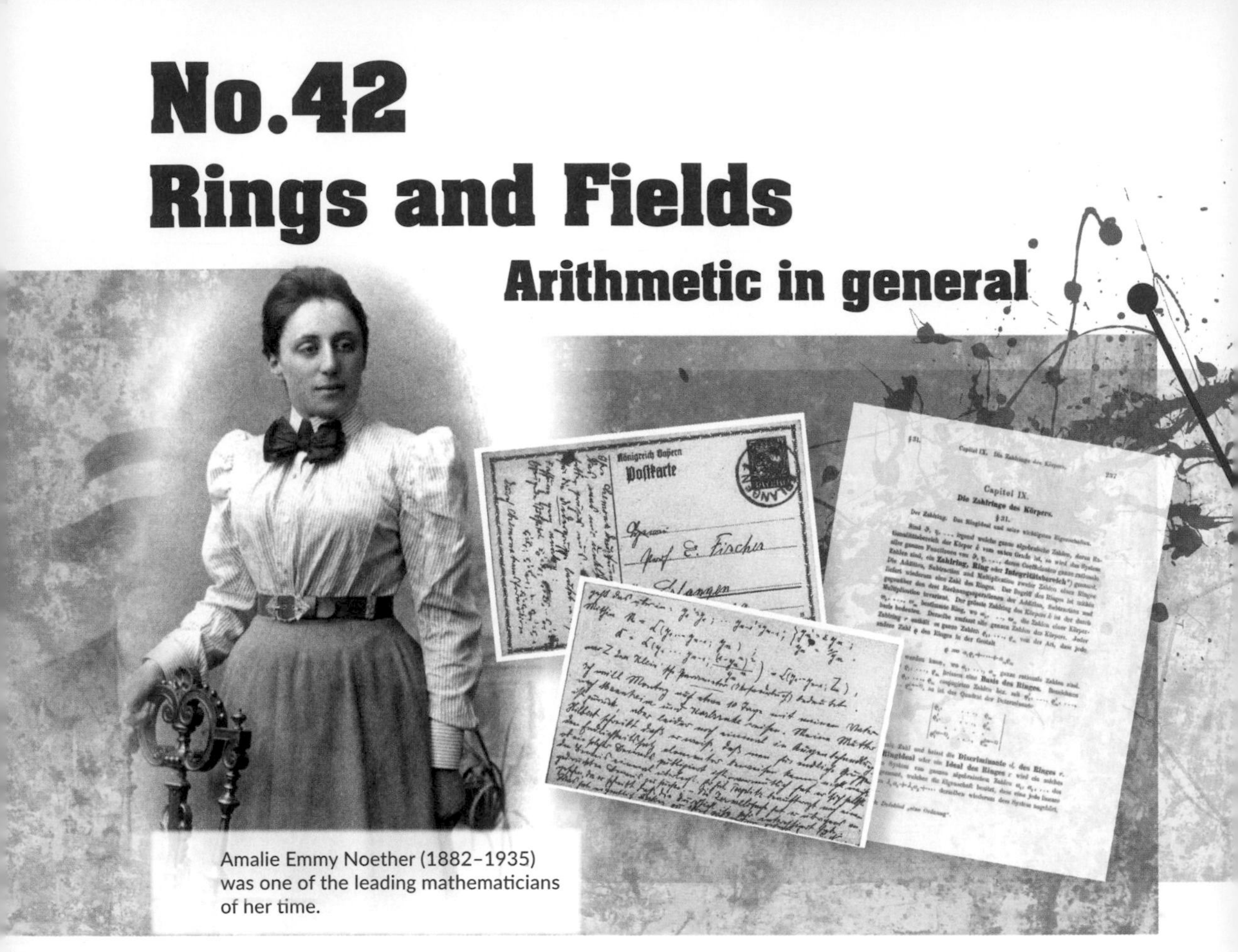

Amalie Emmy Noether (1882–1935) was one of the leading mathematicians of her time.

1/ Helicopter view: Numbers can seem messy. Positive and negative integers, fractions, infinite decimals, complex numbers—the more of them we meet, the more confusing it gets. Just what *is* a number anyway? This is more of a philosophical question than a scientific one.

Perhaps to be a number means to be part of a system where arithmetic makes sense. Such a system would be characterized by its structure and relationships rather than the inner nature of its elements. **Abstract algebra** to the rescue!

Rings are a wide class of structured **sets** that have two **binary operations** that "behave like" addition and multiplication of ordinary numbers. *Fields* are rings in which the algebra is especially nice.

Ring structures are found far beyond sets of numbers, however. Recognizing them in polynomials, operators, topological features, and more yields much information and increases the unity of the subject.

Above: Noether developed the theories of rings, fields, and algebras. Noether sometimes used postcards, such as this, to discuss abstract algebra with her colleague, Ernst Fischer.
Right: The corners of a regular heptagon can be given a field structure and thus treated like number. Those of a hexagon or octagon cannot.

2/ Shortcut: A *ring* is a set with two **binary operations**. One forms a **commutative group** and is thought of as "addition." The other is only required to have an identity and to **distribute** over the first. This is "multiplication."

A *field* is a ring in which multiplication is also a commutative group (ignoring zero). Intuitively, this means division always works in a field. The rational and real numbers both have the field structure, but the integers do not.

See also//

7 Set Theory, p.18

35 Abstract Algebra, p.74

36 Binary Operations, p.76

37 Associative, Commutative, Distributive, p.78

38 Groups, p.80

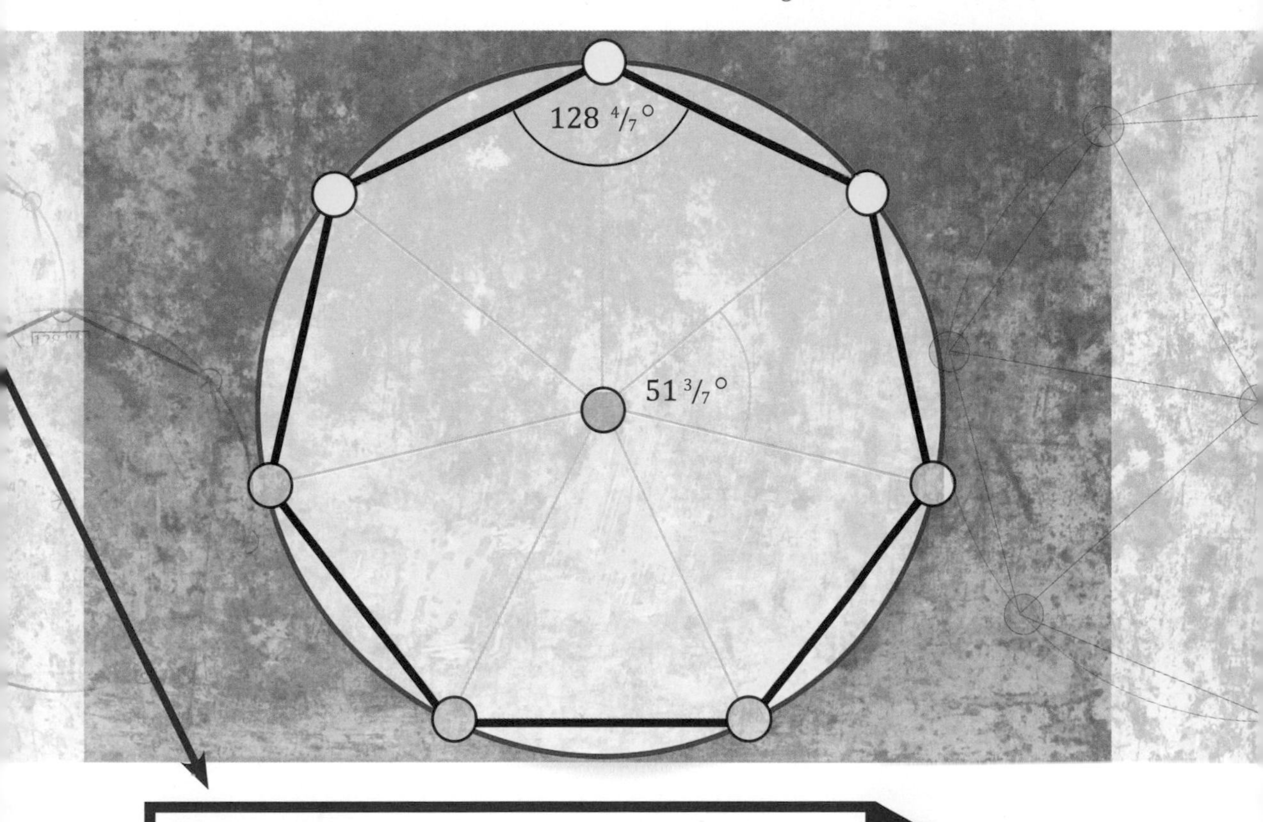

3/ Hack: Rings are structured sets where you can do something that looks like arithmetic. Fields are especially nice rings.

The abstractness of these definitions led to breakthroughs in understanding numbers as well as having far wider applications.

No.43 Galois Theory

Putting fields under the microscope

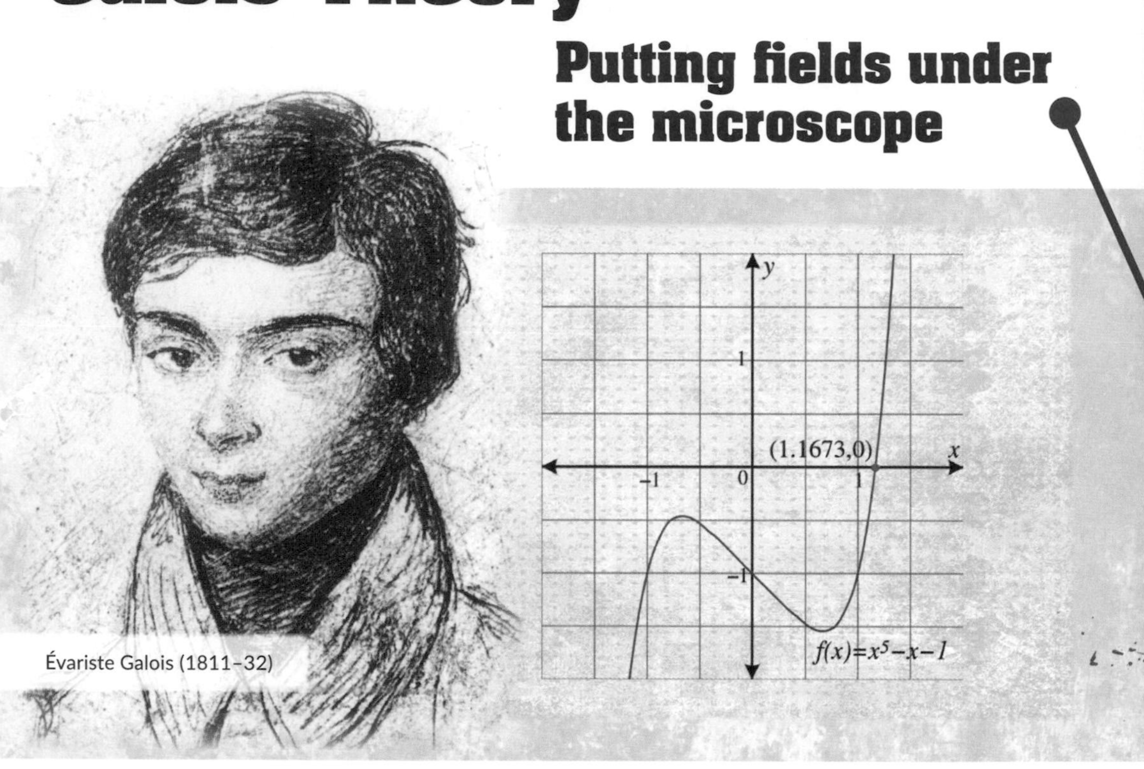

Évariste Galois (1811–32)

1/ Helicopter view: When a number system does not include a number that we need, it may seem excessive to go to a new one that includes a vast quantity of extras we do not need. For example, if we are quite happy working with **rational numbers**, except we want to solve $x^2 - 2 = 0$, why not just add a single number, $\sqrt{2}$, to the rational numbers?

We need to do a little more if we want our number system still to be a **field**; we need to include all the ways of adding and multiplying the new number by any rational number. The result is called a *field extension*. This point of view can bring a specific problem into sharp focus.

In particular, we might be drawn to study **maps** from the field extension to itself that leave everything unchanged except perhaps permuting some of the things we added. These *field automorphisms* form a **group**—the Galois group—whose structure contains information about the extension itself.

Above: The polynomial x^5-x-1 is of order 5, but its graph only crosses the x-axis at one point.
Right: This group of field automorphisms has the same structure as the symmetry group of an equilateral triangle.

2/ Shortcut: *Galois theory* is a general approach that arises from applying symmetries (that is, **group** theory) to field extensions. The details are somewhat complicated but the results can be dramatic. For example, basic Galois theory proves that two ancient problems (doubling the cube and trisecting the angle using only **Euclidean** methods) are *impossible* to solve. Before this was discovered, many great thinkers spent a huge amount of time and effort grappling with them.

See also//
9 Maps, p.22
21 Rational Numbers, p.46
38 Groups, p.80
42 Rings and Fields, p.88
59 Euclidean Spaces, p.122

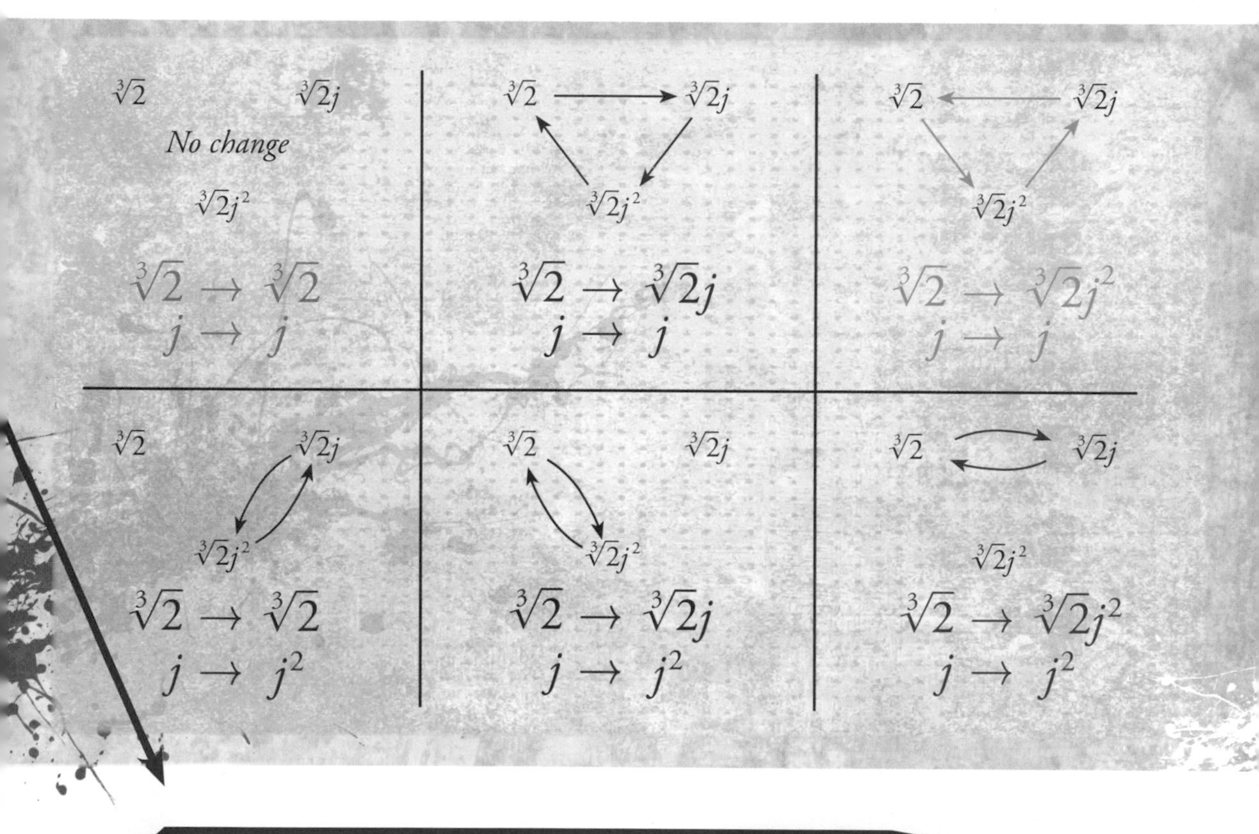

3/ Hack: A field can be extended by adding extra elements; every extension is related to the original field by its Galois group.

Galois theory provides a precise way of probing field extensions and extracting information from them.

No.44
Diophantine Equations
Undecidable problems with whole numbers

Pythagoras (*c.*570–495 B.C.E.)

1/ Helicopter view: A *Diophantine equation* is a **polynomial** equation whose coefficients are all integers, where we want to find integer solutions. For example, $x^2 + 6x - 16 = 0$ is true when $x = 2$ or when $x = -8$ (check for yourself!). On the other hand, $x^2 - 2 = 0$ has no integer solutions.

Puzzles about these equations are quite ancient. The school of Pythagoras, for instance, looked for solutions to $x^2 + y^2 = z^2$, which is related to the famous theorem about triangles. You might know that $x = 3$, $y = 4$, $z = 5$ is one solution. There are an infinite number of others; at least three methods for finding them were known by 350 B.C.E.

In 1900 David Hilbert described twenty-three unsolved problems he believed were important for the next century of mathematics. The tenth asked for an algorithm (in modern terms, a computer program) for discovering whether any given Diophantine equation has solutions. Seventy years later, Yuri Matyiasevich proved that no such algorithm can ever exist, building on the methods used in **Gödel's First Incompleteness Theorem**.

Above: Pythagoras and his school considered whole numbers to be magical and placed great stock in problems concerning them. *Right:* The polynomial graphed on the left has integer solutions. The one on the right does not.

2/Shortcut: What makes *Diophantine equations* difficult is that the integers are a **ring** in which the operations of multiplication and raising to a power do not usually have inverses.

Using **rational numbers**, we can solve $5x = 3$ by dividing both sides by 5 to give $x = 3/5$, but $3/5$ is not an integer, so this is ruled out. Similarly, if $x^2 = 12$, $x = \sqrt{12}$, but this is only available in the **real numbers**.

It gets much harder to "see" whether a solution exists when dealing with a complicated equation that may allow many routes to a solution.

See also//

6 Gödel's Incompleteness Theorems, p.16

20 Negative Numbers, p.44

21 Rational Numbers, p.46

23 Polynomials, p.50

26 Real Numbers, p.56

42 Rings and Fields, p.88

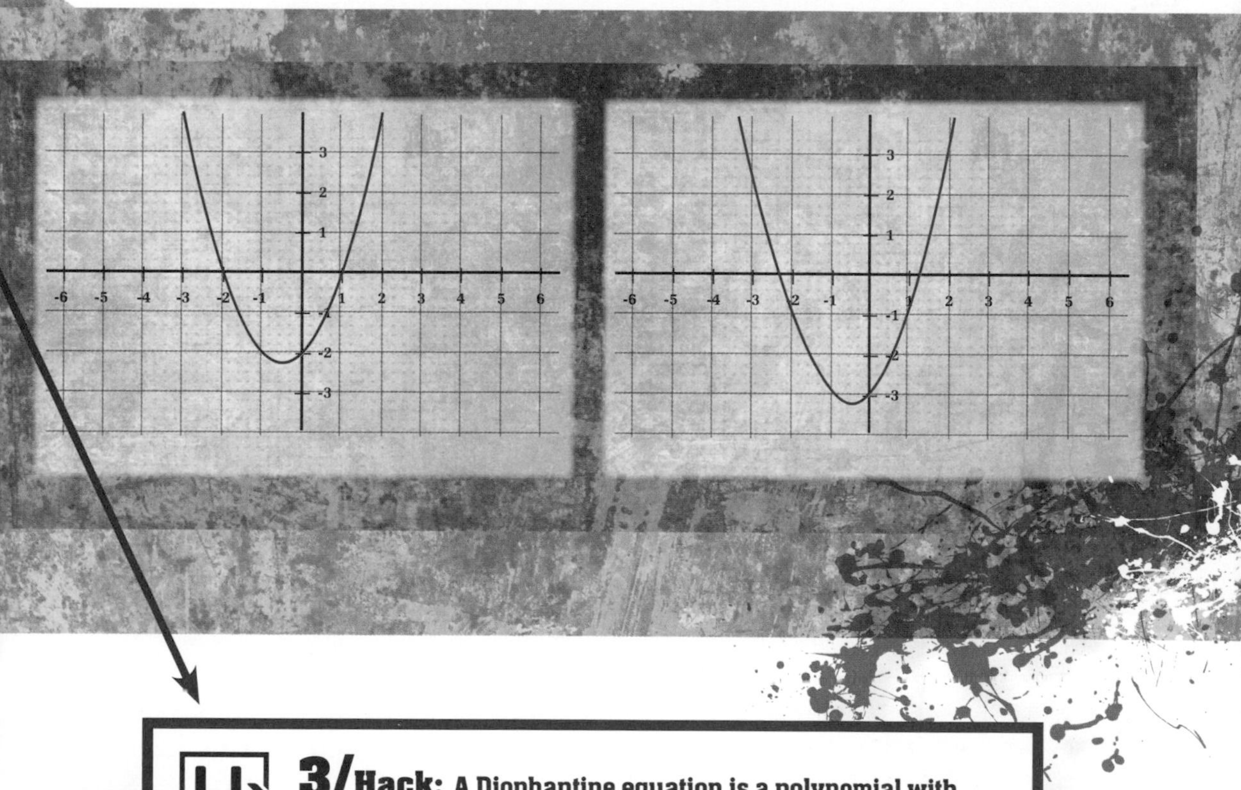

3/Hack: A Diophantine equation is a polynomial with integer coefficients that requires integer solutions. Some have solutions, others do not, and we can't easily tell which.

We still do not know the full picture when a number system other than the integers is used.

No.45
Fermat's Last Theorem

Solving a problem with no solutions

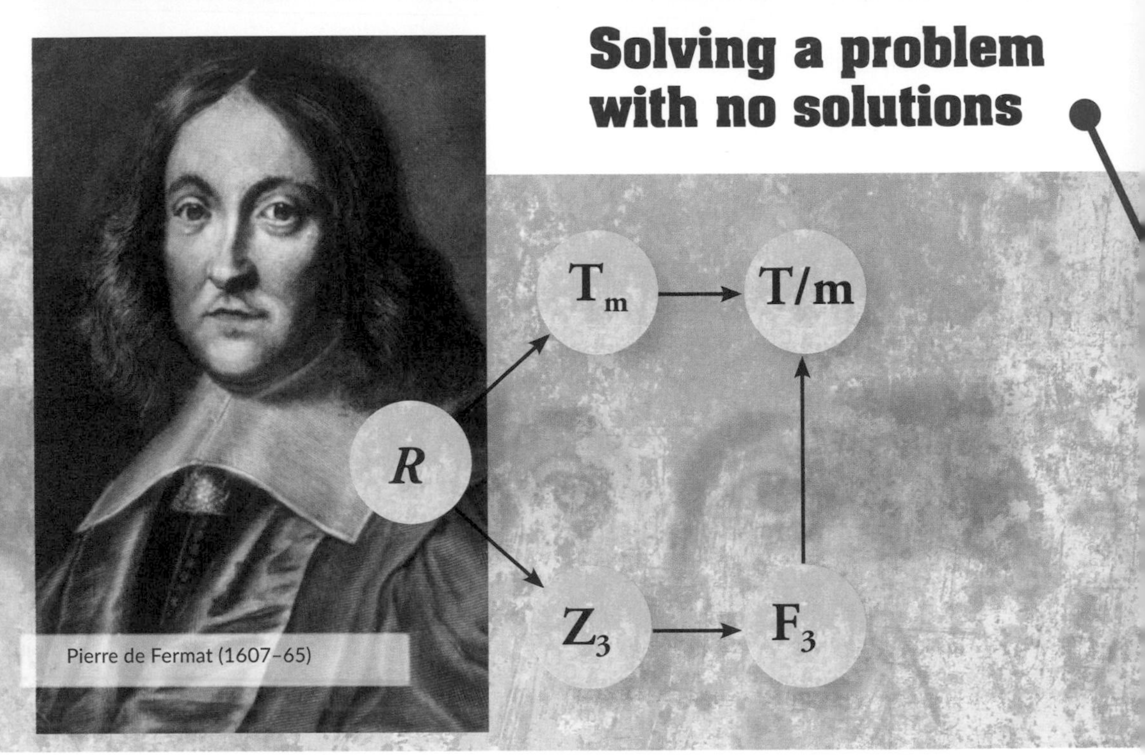

Pierre de Fermat (1607–65)

1/ Helicopter view: One family of **Diophantine equations** looks like this: $x^n + y^n = z^n$, for some **natural number** n. When $n = 0$ this is trivial: it becomes $1 + 1 = 1$, which is false. When $n = 1$ it is almost as easy: $x + y = z$ has an infinite number of solutions, which are very easy to find. When $n = 2$ we have: $x^2 + y^2 = z^2$, a harder problem, but it had been long known to have an infinite number of solutions.

What about higher values of n? *Fermat's Last Theorem* states there are never any solutions at all for any of them! This is pretty surprising, given that the first few were so easy.

Although Pierre de Fermat claimed, in a famous marginal note, to have found a proof in 1637, it seems extremely unlikely that he had one. The result was not proved until 1994, and to do it Andrew Wiles used advanced techniques that would have been not only unavailable to Fermat but probably incomprehensible to him.

Above: Andrew Wiles's proof uses modular forms, a new and advanced branch of mathematics unknown to Fermat.

2/Shortcut: The truth of *Fermat's Last Theorem* had no immediate applications or significance for other parts of mathematics, and it was always a classic pure-mathematical puzzle. Yet many useful techniques and ideas were developed in search of a solution.

In the decade after Wiles's proof was published, other results emerged based on his methods. One, the Modularity Theorem, reveals deep connections between topology and number theory that can be seen as part of the far-reaching Langlands Program.

See also//
14 Natural Numbers, p.32
43 Galois Theory, p.90
44 Diophantine Equations, p.92

Andrew Wiles (b.1953)

3/Hack: The theorem may be a mere puzzle, but the proof developed new ideas and techniques that are being applied much more widely.

$x^n + y^n = z^n$ has no nonzero integer solutions when $n > 2$.

No.46
The Unsolvable Quintic

One, two, three, four but not five

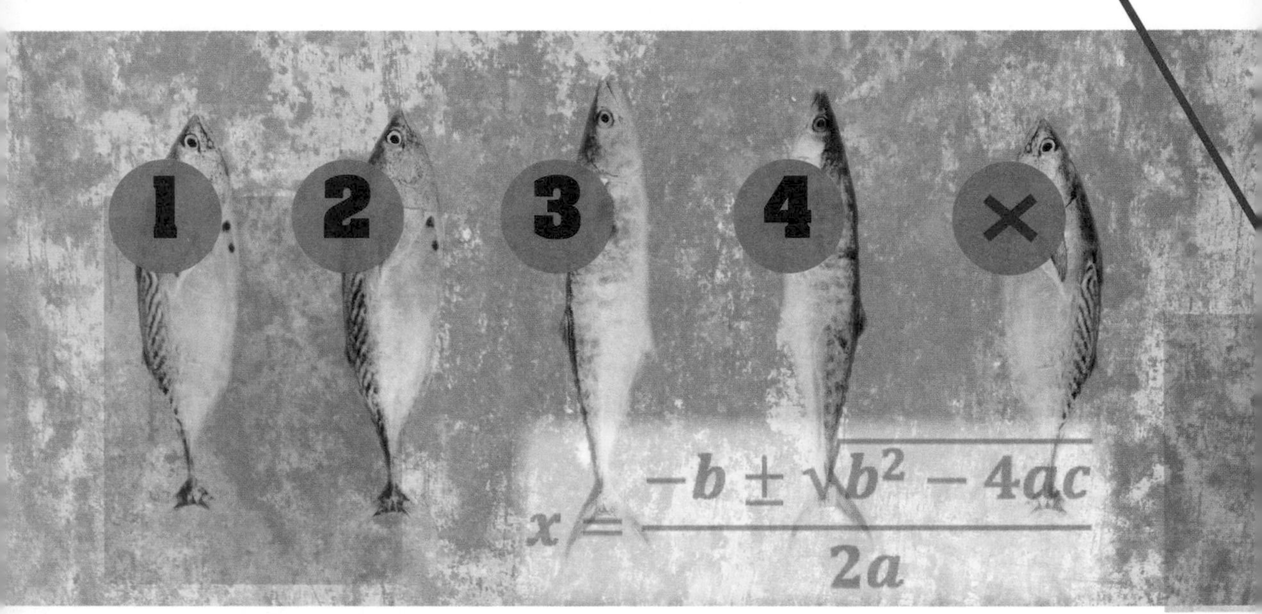

1/ Helicopter view: Many schoolchildren learn how to solve *linear equations* of the form $bx + a = 0$ where a and b are numbers. For example, $3x - 12 = 0$ is only true when $x = 4$, so that is the solution.

Later we study *quadratic equations*, which add an x^2 term: $cx^2 + bx + a = 0$. For example, $x^2 + 2x - 15 = 0$ is true when $x = 3$ or when $x = -5$, so these are the solutions. A single formula works for all of them, and it uses only addition, subtraction, multiplication, division, and a square root sign.

In the 16th century a similarly general formula was found for cubic equations, which add a dx^3 term. It's complicated, but again it only involves arithmetic and roots. Around the same time Ludovico Ferrari found an even more complicated equation for quartic equations, which look like $ex^4 + dx^3 + cx^2 + bx + a = 0$, again with the same basic components.

What about quintic equations, which add an fx^5 term?

Above: The formula for solving quadratic equations is taught to schoolchildren. *Right:* The formula for cubic equations is much more complicated.

2/Shortcut: It is natural to expect an equation can be found for quintics, too, but in 1825 Niels Abel showed this is impossible.

The sudden breaking of a pattern like this always piques mathematicians' curiosity. This led Évariste Galois to develop the **abstract algebraic** methods that laid the foundations of **Galois Theory**.

Showing that something is impossible in principle, no matter how hard we try, is usually rather difficult and often requires, as in this case, the development of new mathematical ideas.

See also//

23 Polynomials, p.50

35 Abstract Algebra, p.74

43 Galois Theory, p.90

44 Diophantine Equations, p.92

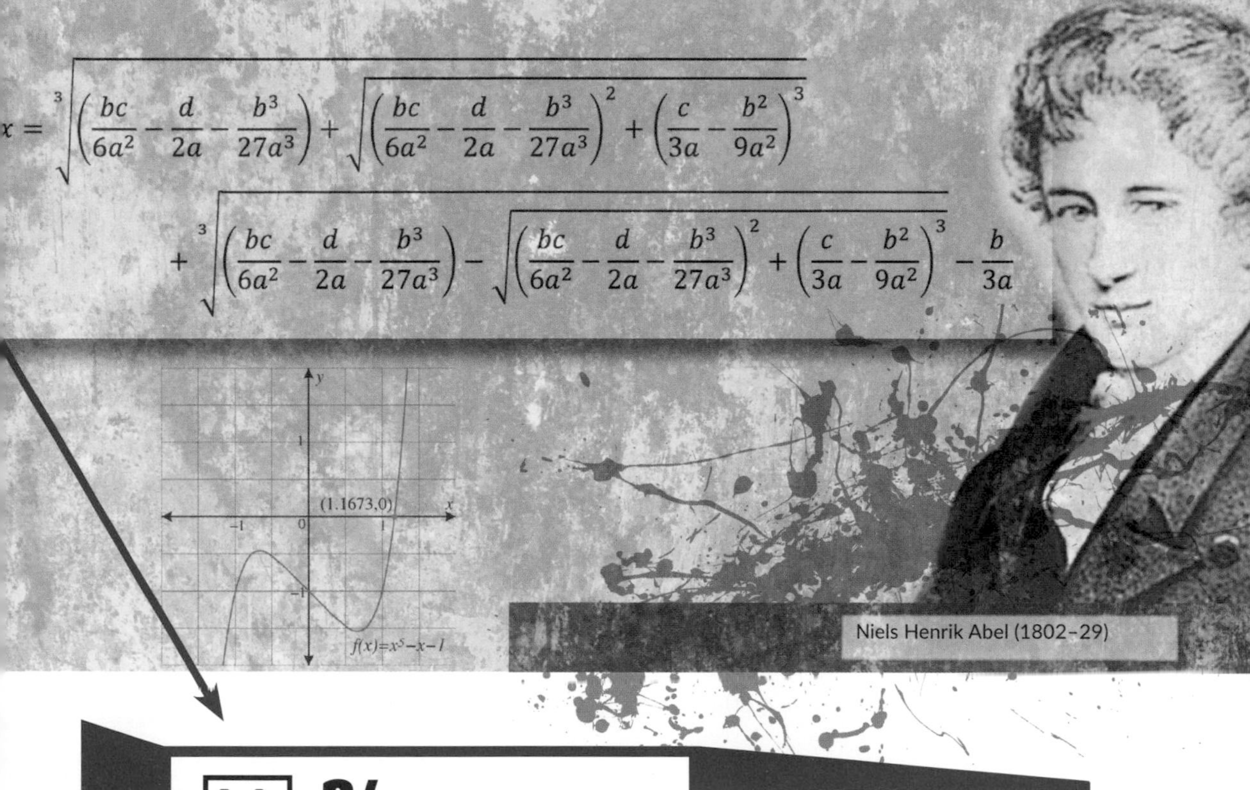

Niels Henrik Abel (1802–29)

3/Hack: There is one quadratic formula, one cubic, and one quartic. However, no algebraic formula can ever exist that solves all quintic equations.

The pattern we expect is not there and the real pattern, revealed by Galois theory, is much more subtle.

No.47
The Riemann Hypothesis
Hunting for zeroes

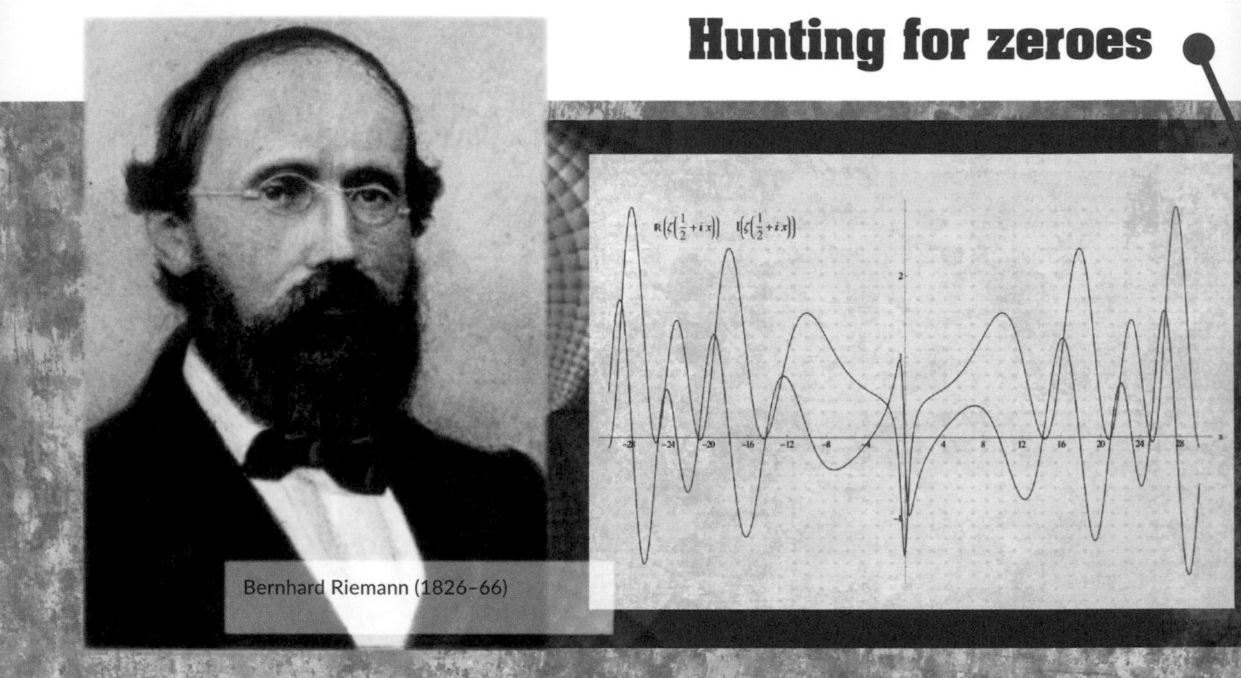

Bernhard Riemann (1826–66)

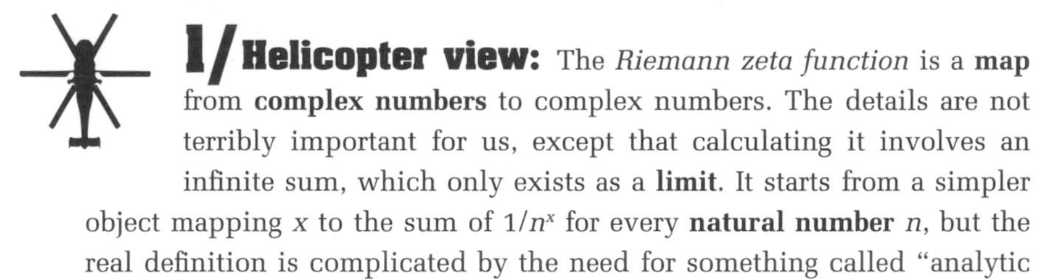

1/ Helicopter view: The *Riemann zeta function* is a **map** from **complex numbers** to complex numbers. The details are not terribly important for us, except that calculating it involves an infinite sum, which only exists as a **limit**. It starts from a simpler object mapping x to the sum of $1/n^x$ for every **natural number** n, but the real definition is complicated by the need for something called "analytic continuation."

A natural question to ask is which numbers, if any, the zeta function maps to zero. It turns out that this happens to every negative even integer—that is a consequence of the way analytic continuation works. These values are somewhat obvious, so they are called *trivial zeroes*.

We also know of many *nontrivial zeroes* of the zeta function, but all of them are **complex numbers** of the form ½ + xi. The *Riemann Hypothesis* states that there are no other, unexpected, zeroes anywhere. All are in one of these two categories.

All nontrivial zeroes of the zeta function are known to lie within the critical strip but do they all lie on the critical line?

2/Shortcut:

The *Riemann Hypothesis* has (at the time of writing) been found to be true by computer for the first 1,013 nontrivial zeroes. If even one single exception were found, the hypothesis would be disproved. If it is true, as most people seem to believe, such methods will never prove it.

There is a deep connection between the distribution of zeroes of the zeta function and the distribution of **prime numbers**. The hope is that a proof will help us understand primes better.

See also//
4 Limits, p.12
9 Maps, p.22
14 Natural Numbers, p.32
17 Prime Numbers, p.38
20 Negative Numbers, p.44
33 Complex Numbers, p.70

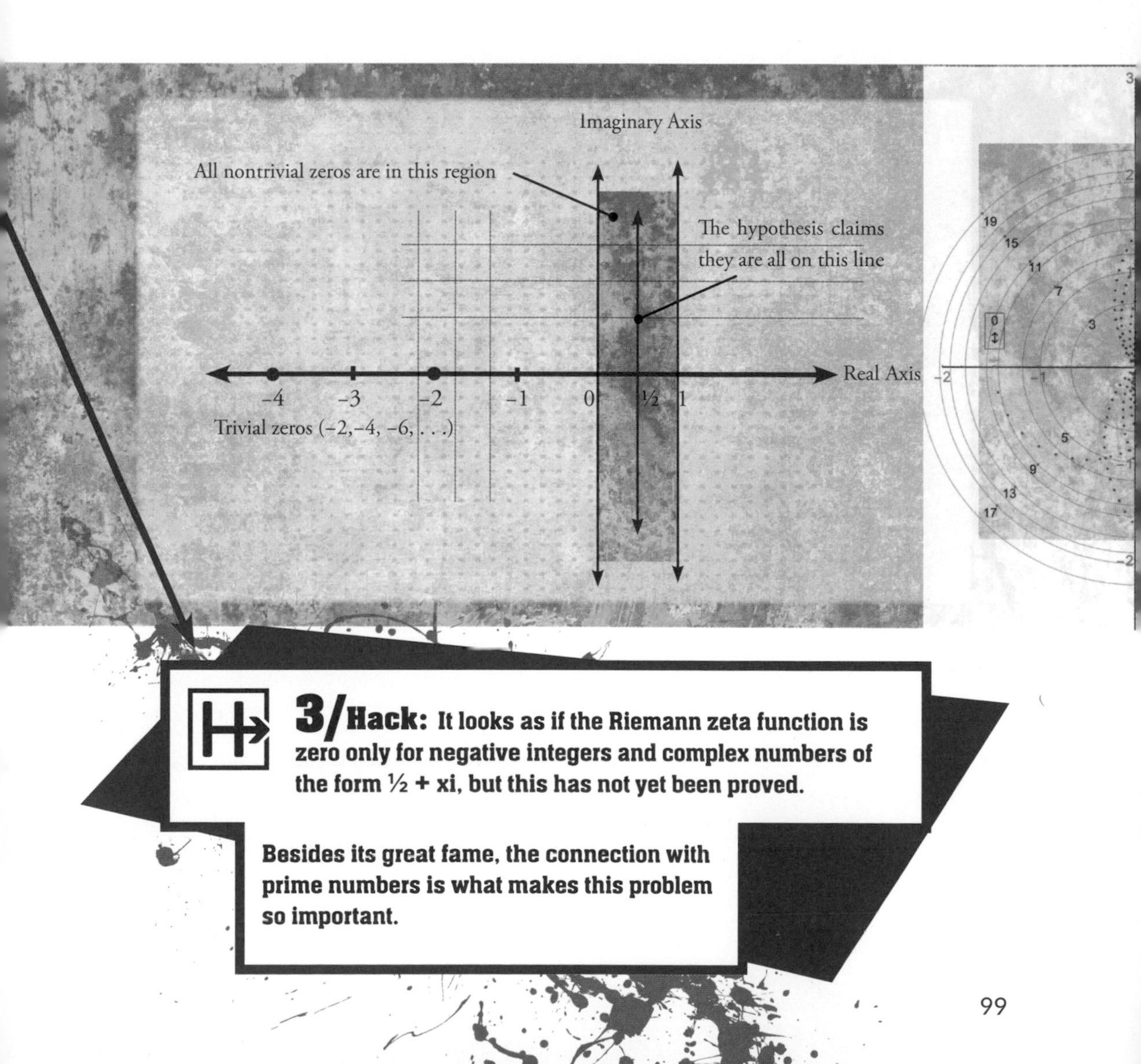

3/Hack:

It looks as if the Riemann zeta function is zero only for negative integers and complex numbers of the form ½ + xi, but this has not yet been proved.

Besides its great fame, the connection with prime numbers is what makes this problem so important.

No.48

Order Theory

More or less structure

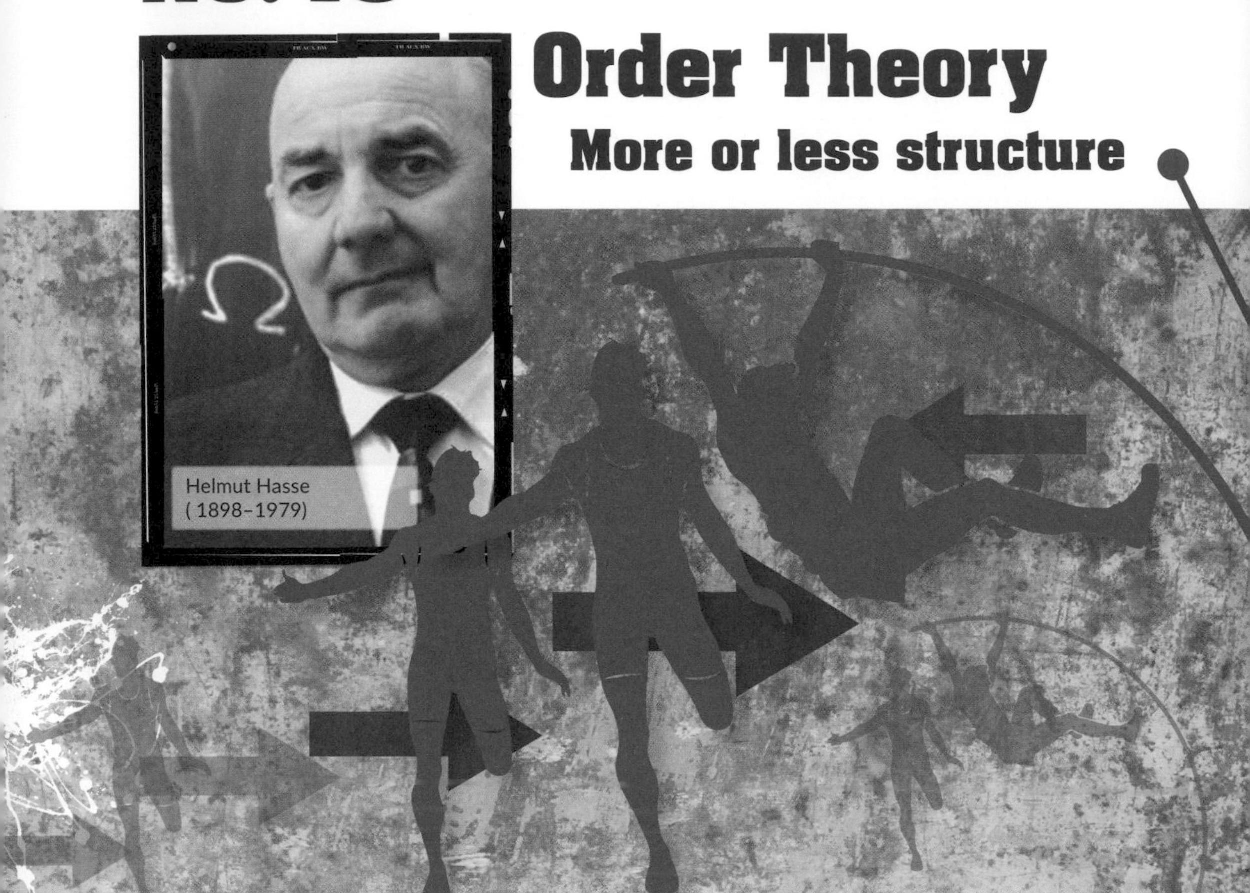

1/ Helicopter view: A **set** is said to be "ordered" if it makes sense to pick two objects and ask whether one is less than another.

Any set of natural numbers, for example, is ordered because for any pair we can always say which is less than the other. A set of products for sale in a shop might be ordered by how much they cost. Some pairs we pick will both be the same price, but that is OK. The answer to "is this less than that?" is "no" in both cases.

Now consider the set of competitors in the Olympics. We say "A is less than B" if B beat A in an event. If we pick two 100m sprinters, it may be that $A < B$ or $B < A$. But what about one sprinter and one pole vaulter? In this case the comparison does not make sense. Here we have only a *partial order*.

Above: Ordering things can be more complicated than simply ranking them in a single sequence. *Right:* Hasse diagrams place "greater" things higher up; the divisors of 120 and these sets ordered by containment have the same order structure.

2/Shortcut: Mathematical objects can often be ordered in natural and obvious ways. At other times we impose a weird-looking order because it helps us see things in a new light.

As with all **algebraic** structures, we use formal definitions that we can develop a lot of abstract theory about. That makes them very useful when we apply them.

An order on a **set** is rather simple; richer structures such as lattices and (sometimes) *exact sequences* can be built out of them.

See also//
7 Set Theory, p.18
35 Abstract Algebra, p.74
49 Homological Algebra, p.102
93 Combinatorics, p.190

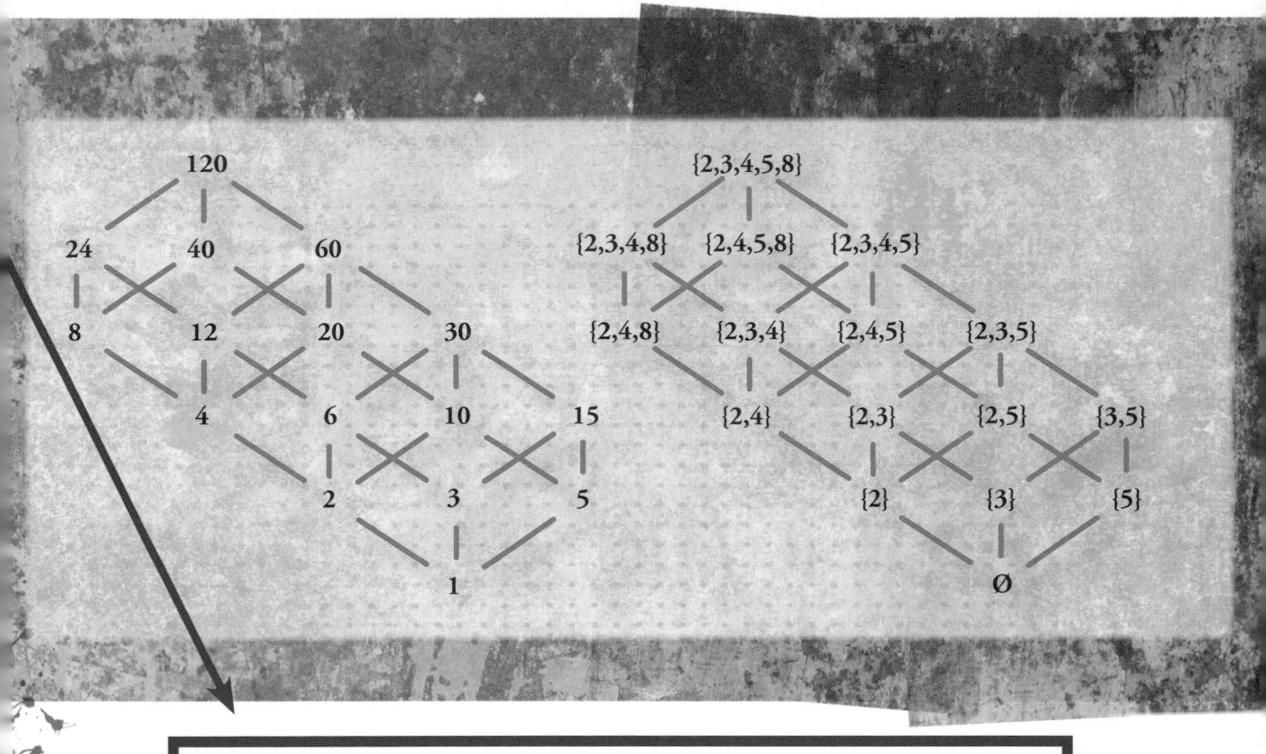

3/Hack: Order is an abstract structure that allows us to compare pairs of objects and say which is "bigger," but leaves us free to define "bigger" in any way we like.

Ordered sets have many applications, including deep ones in logic, computing, combinatorics, and topology.

No.49
Homological Algebra
Probing sequences of maps

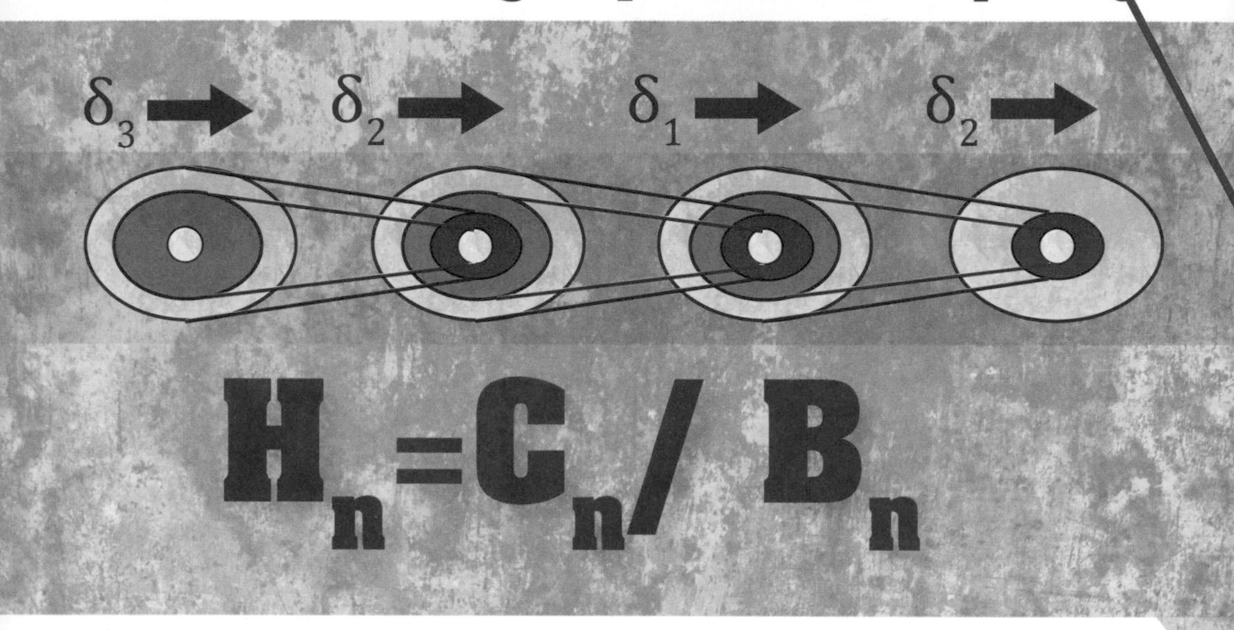

1/ Helicopter view: **Maps** between **sets** can be *composed*, that is, carried out one after another. For example, I can map the set of diners to the set of meals on the menu, then map the meals to a set of prices to find out how much each diner spent.

In mathematical settings, chaining maps together like this quite often gives rise to an object called a *chain complex* in which, very roughly, passing along any two consecutive maps collapses everything down to a single point.

Even more roughly, homology measures how efficiently this disappearing is achieved. If it is as efficient as possible, the chain complex is called an *exact sequence*.

We know a lot of very general things about exact sequences. Thus, if you can prove that some situation you are dealing with involves an exact sequence, you get a lot of extra information for free. What is more, *homological algebra* gives you powerful tools for digging deeper.

Above: Homological algebra, established on a rigorous footing by Samuel Eilenberg and Norman Steenrod in the 1940s, studies structures that look rather special but that turn up in many different mathematical contexts.
Right: Tor, a sort of mirror image of the tensor product, is part of the toolkit for probing exact sequences.

2/ Shortcut: Exact sequences, and complexes that fail to be exact are surprisingly common in abstract mathematics. *Homological algebra* arose as a way to study them. It has since become an almost universal tool in topology, geometry, and algebra.

The subject has deep roots in **algebraic topology**, in which homology is a crucial tool. It is also central to the study of **manifolds**, and thus has physics applications, too.

Today it is developed as part of **category** theory; many of its arguments work by drawing diagrams.

See also//
7 Set Theory, p.18
9 Maps, p.22
13 Categories, p.30
60 Manifolds, p.124
70 Topology, p.144
84 Algebraic Topology, p.172

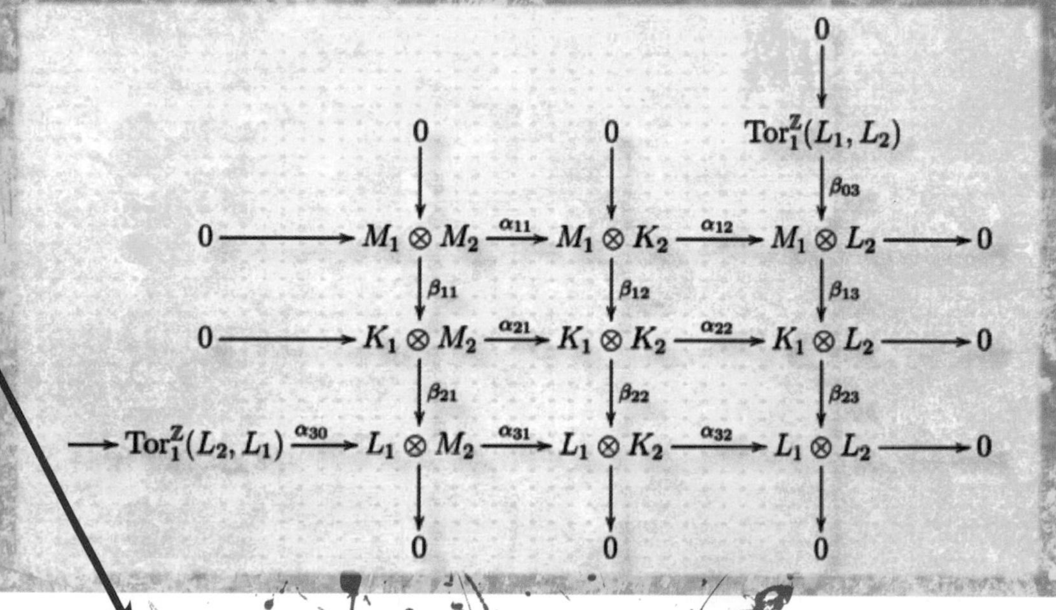

3/ Hack: It is very common to find maps chained together in a sequence. Homological algebra studies properties of such sequences that are far from obvious.

Homology began as an application of category theory to topology, but has now spread into many other fields.

No.50 The Derivative

Going off at a tangent

1/ Helicopter view: Many physical situations can be modeled by a **map** from the set of **real numbers** to itself, which we will call a *function*. We sometimes think of the function f as representing some quantity (the *codomain*) that changes over time (the *domain*). A **polynomial** is an example of a function.

The *derivative* of f, written as f', gives us a number that tells us how fast f is changing at each moment in time. This f' is a new function derived from f (hence the name).

Alternatively, the rate of change of f at a point can be thought of as the tangent to the graph of f there (you can think of this as a straight line that "just touches" the graph). The idea of a *derivative* first emerged from the application of this geometric picture to problems about velocities in physics.

If a curve represents a motion happening over time, the tangent line at a point describes something like the "instantaneous velocity" of the motion.

Changes that happen over time are easy to picture, but derivatives are not limited to that; the idea of "change" extends to other situations, too.

2/ Shortcut: The *derivative* of a function gives us information about how it is changing at each point. Consider the function $f(t) = t^2$, which maps each number to its square. The derivative can be easily calculated as $f'(t) = 2t$.

Suppose we drop a ball from a high window. The time t is measured in seconds, and $f(t) = t^2$ tells us the number of meters the ball has fallen at time t. Then 1 second after we release the ball, the ball's position is $f(1) = 1$ meter away, but it is falling at $f'(1) = 2$ meters per second.

See also//
9 Maps, p.22
26 Real Numbers, p.56
53 The Fundamental Theorem of Calculus, p.110

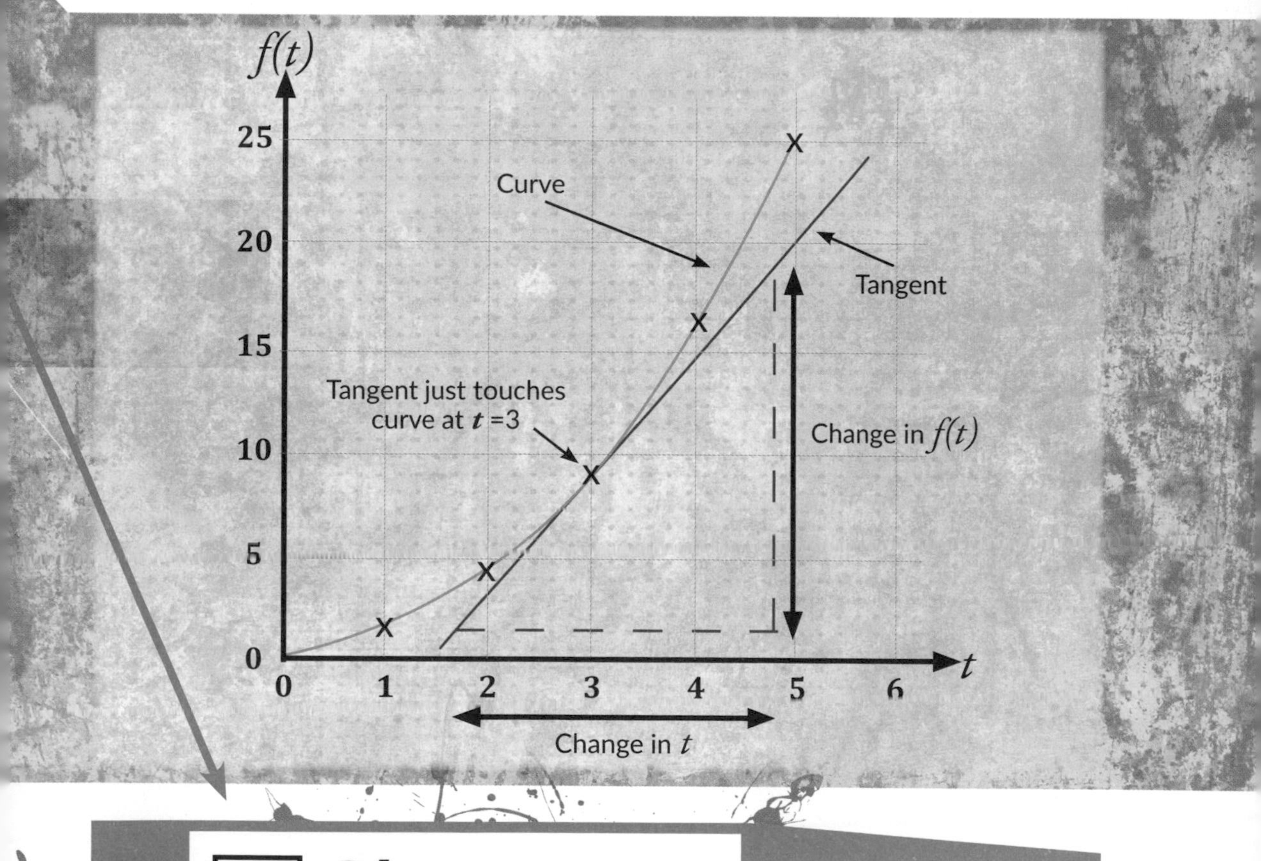

3/ Hack: The derivative of a function is another function that indicateshow it is changing at each point in its domain.

The derivative captures the idea of rate of change.

No.51
Taylor Series
Approximating awkward functions

Brook Taylor (1685–1731) published his "incremental method" in 1715.

1/Helicopter view: **Polynomials** are very nice, well understood functions, but many important functions are much harder to deal with. Examples include **logarithms**, exponentials like $f(x) = 2^x$, trigonometric functions like sine, cosine, and tangent, and many others.

In 1715 Brook Taylor found a way to approximate these awkward functions with sums that look like polynomials except that they usually have an infinite number of terms.

They are determined by repeatedly taking the **derivative** of the function you are trying to approximate. We can make the approximation as close as we like by calculating more and more terms from the infinite sum because, in the **limit**, the sum is equal to the function it approximates.

For some functions (called *entire functions*), the *Taylor series* at any single point gives a good approximation at every point. For others, such as logarithms, the approximation only works for regions close to the point where it is calculated.

Right: The sine function is not a polynomial, but its Taylor series at 0 is an ever-growing polynomial that provides an ever-improving approximation of it there.

2/ Shortcut:

Taylor series are convenient because it's usually easy to calculate the **derivatives** we need. The individual parts of the sum are pretty simple, and we can take as many as we need for practical purposes when making an approximation.

At the beginning of the 19th century the idea of approximating a function by an infinite sum of simpler things was extended by Joseph Fourier, who was then able to conquer previously intractable problems in the theories of heat and waves.

See also//
4 Limits, p.12
23 Polynomials, p.50
24 Logarithms, p.52
50 The Derivative, p.104

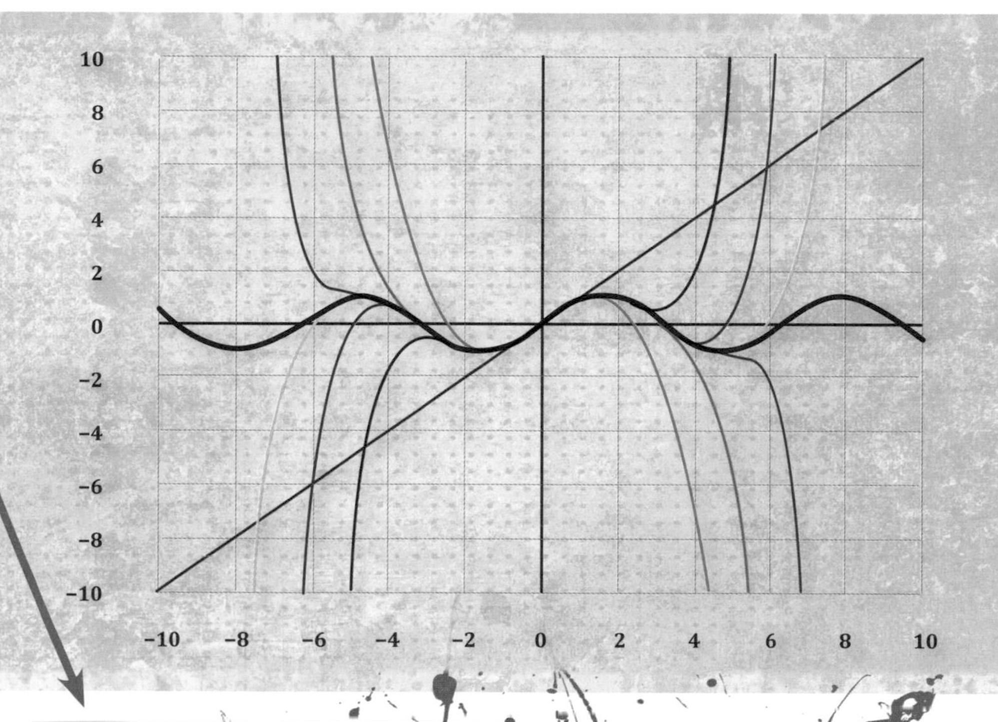

3/ Hack:

A function's Taylor series, if it has one, is an infinite sum that is equal to the function in the limit, at least around the point where it is calculated.

Infinite sums are practical because we can stop calculating when we have as much accuracy as we need.

No.52 The Integral

Squaring circles and other curves

Archimedes (*c*.287–212 B.C.E.) and others found ways to approximate curved areas by cutting them into ever smaller parts.

1/ Helicopter view: Area is the two-dimensional equivalent of length; it measures how much space a shape takes up. Euclid, around 300 B.C.E., showed how to calculate the areas of straight-sided shapes by turning them into squares without changing their areas.

However, **Euclidean** geometry could not cope with shapes with curved boundaries. Their areas could only be approximated, and where they could be found it was by special arguments that did not generalize. Some of the ideas they used, however, were revived and explored in the 17th century, leading gradually to the definition of what we now call the *integral*.

There are various ways to define integrals but most involve cutting up the region whose area you want to find into smaller and smaller parts using straight lines. Each part may still have some curved parts on its boundary, but as we repeat the process these become smaller and less significant in comparison with the straight parts of the boundary.

Right: The integral of a curve arises from cutting the area under it into ever smaller strips.

2/ Shortcut: The key idea is to take this process to an infinite **limit**, in which any effect curvature might have on the area vanishes. For example, a Riemann integral cuts the shape into long, thin strips. As the strips get thinner the curved parts become vanishingly small.

The difficulty is showing that the area does not itself simply become zero when this happens. In the 19th century much more careful definitions put the whole of **calculus**, including *integrals*, on a more solid foundation.

See also//
4 Limits, p.12
53 The Fundamental Theorem of Calculus, p.110
59 Euclidean Spaces, p.122

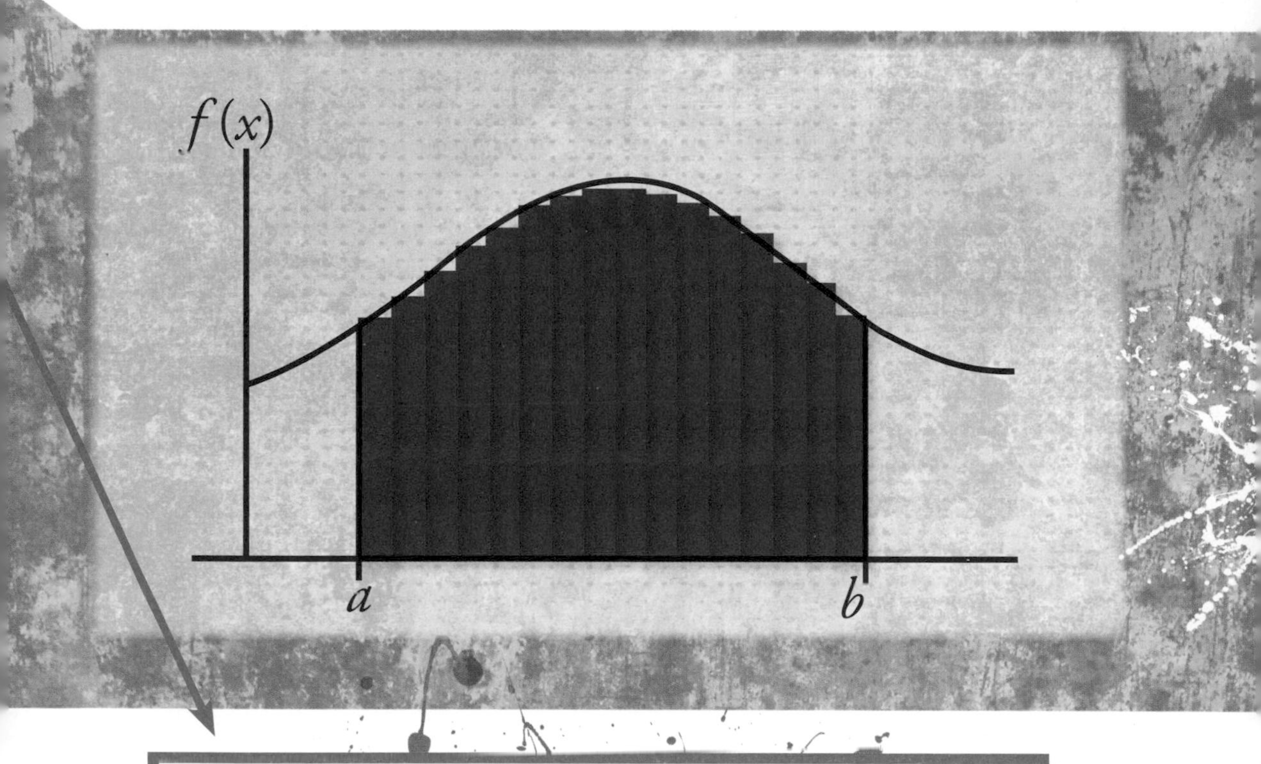

3/ Hack: Cut up a region of space into pieces with straight boundaries, then add them to get an approximate area. The smaller the pieces, the better the approximation.

An integral is the limit of an infinite number of approximations to represent an area.

No.53 The Fundamental Theorem of Calculus

Integrals and derivatives undo each other

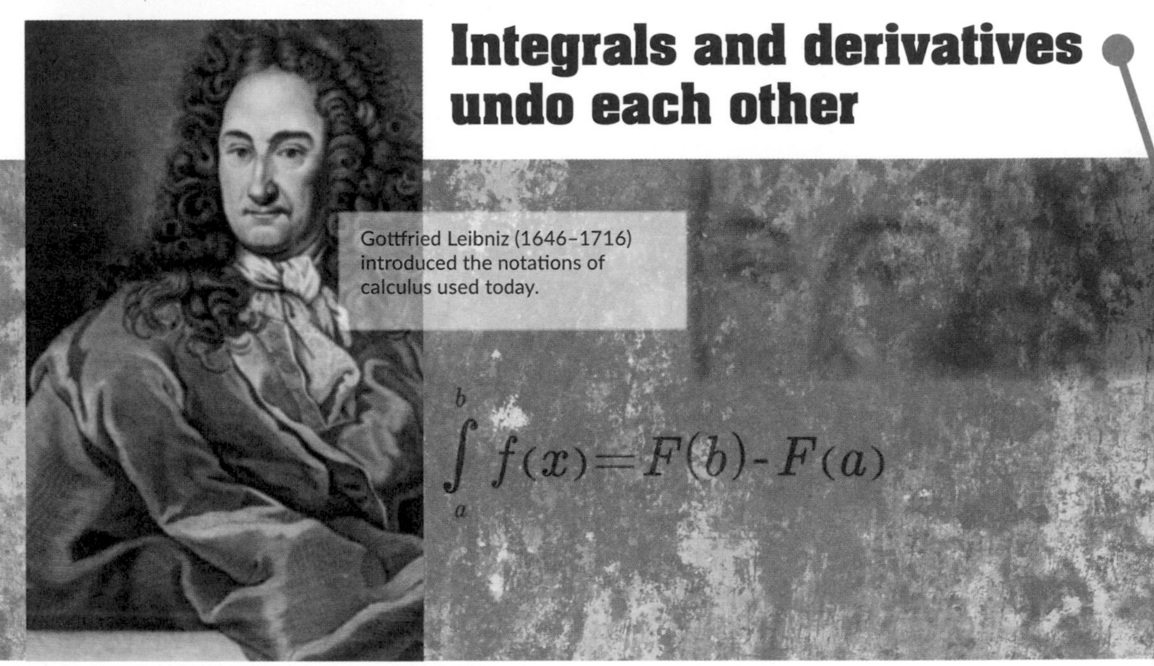

Gottfried Leibniz (1646–1716) introduced the notations of calculus used today.

1/ Helicopter view: Geometrically, **derivatives** were used to find tangents to curves, whereas **integrals** calculate areas. In practical terms, derivatives capture the notion of a rate of change, but integrals do not seem to have much to do with that. The great surprise is that derivatives and integrals are intimately related.

The derivative maps functions to functions. It takes a function and gives us back another one, which tells us how the original function is changing at each point. Integrals can be understood as mapping functions to functions, too; the integral of a function tells us how to calculate the areas of regions that have the graph of the function as part of their boundaries.

The *Fundamental Theorem of Calculus* says that, roughly speaking, if you start with a function f, take its derivative and then integrate the result, you will get f again. In other words, integration and differentiation are **inverses** of each other.

Right: Differentiating f takes us to f' (from left to right); integrating takes us back again.

2/ Shortcut: Suppose f is a function and f' is its **derivative**. Then f' at each point tells us how the function is changing at that point. If we could add up all those tiny changes we should recover the original function f again. This, it turns out, is what the **integral** does.

On their own, derivatives and integrals are handy but rather specialized techniques for solving particular problems. That they are unified in this way makes things *much* more interesting.

See also//
11 Inverses, p.26
50 The Derivative, p.104
52 The Integral, p.108
55 Differential Equations, p.114

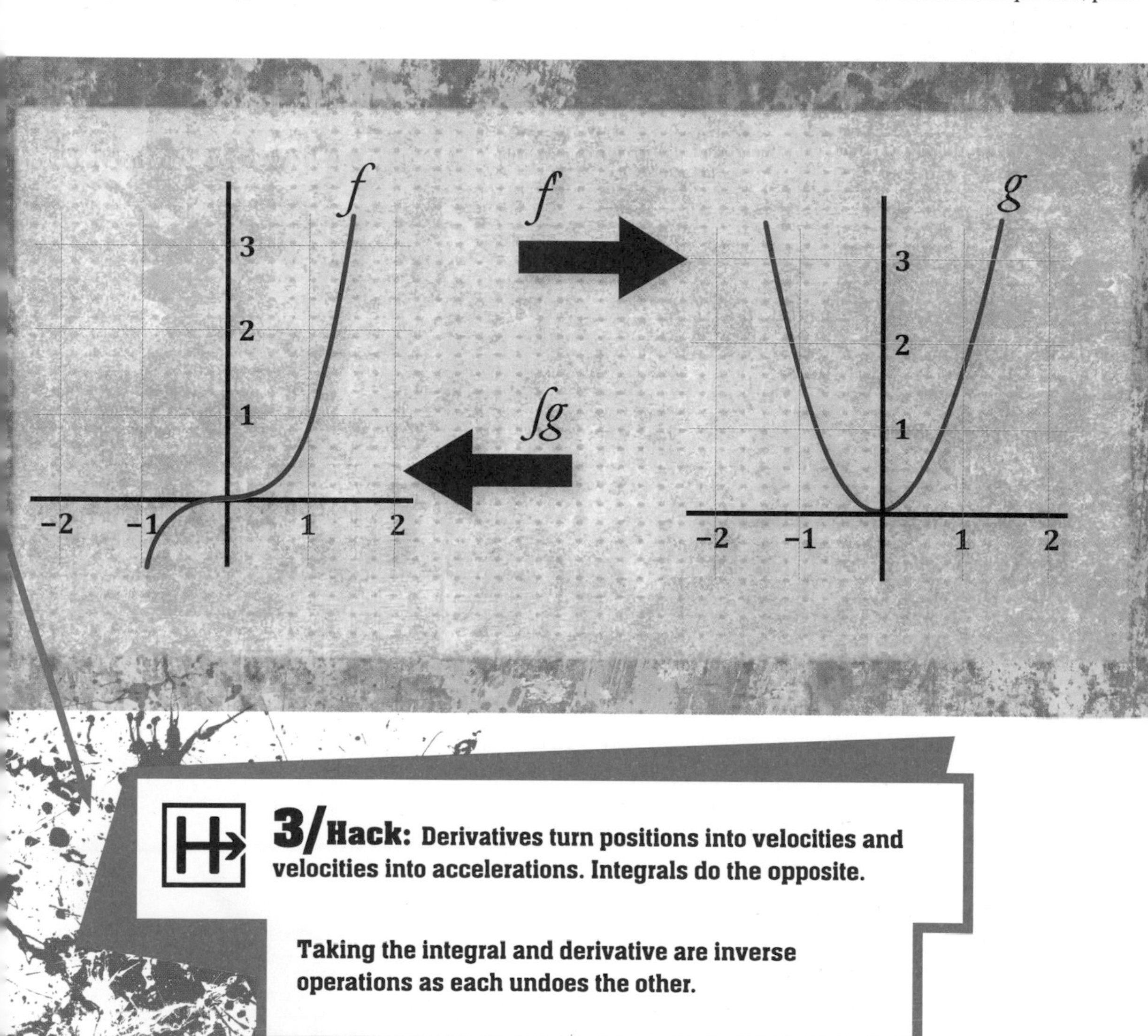

3/ Hack: Derivatives turn positions into velocities and velocities into accelerations. Integrals do the opposite.

Taking the integral and derivative are inverse operations as each undoes the other.

No.54
Pathological Functions
So weird they broke calculus

Karl Weierstrass (1815–97) published his function in 1872.

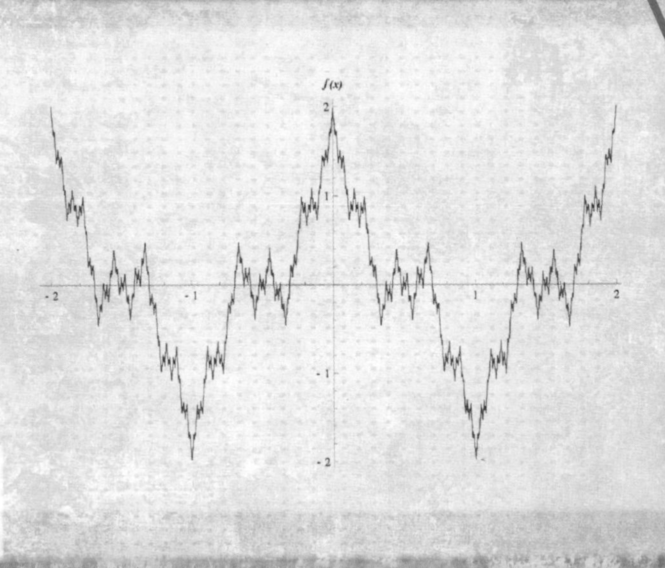

1/Helicopter view: Once the **calculus** came into widespread use it gave rise to many practical applications and new mathematical ideas. As these developed, people began to worry that perhaps the ideas it was based on—infinitesimal quantities and **limits** of infinite processes, in particular—might not be well understood.

That this was indeed the case was dramatically emphasized by the appearance of paradoxes and "pathological" objects that should not exist according to the intuitions and practical considerations that gave rise to them. The vague, intuitive definitions of early calculus allowed these to exist but were not precise enough to tell us how to deal with them.

Such monstrosities have attracted various reactions from mathematicians and philosophers. Some say they simply attest to the power of calculus; others that they show that the underlying logic of the subject is too lax. Such debates continue among philosophers, although most mathematicians are happy to accept the existence of *pathological functions*.

Right: The Dirichlet function is defined at every point on the real number line but is not continuous anywhere.

2/ Shortcut: Karl Weierstrass published the most famous *pathological function* in 1872. It is the limit of a Fourier series (similar to a **Taylor series**), consisting of an infinite superimposition of ever smaller waves. It does not look unusual at first glance.

Intuitively, though, the graph of the function seems to be a continuous line with a corner at every point, which seems geometrically impossible. In calculus terms, it has no **derivative** at any point.

Many pathological functions are, like this one, *fractals*, although that idea was not developed until a century later.

See also//
3 Reductio ad Absurdum, p.10
4 Limits, p.12
50 The Derivative, p.104
51 Taylor Series, p.106
53 The Fundamental Theorem of Calculus, p.110

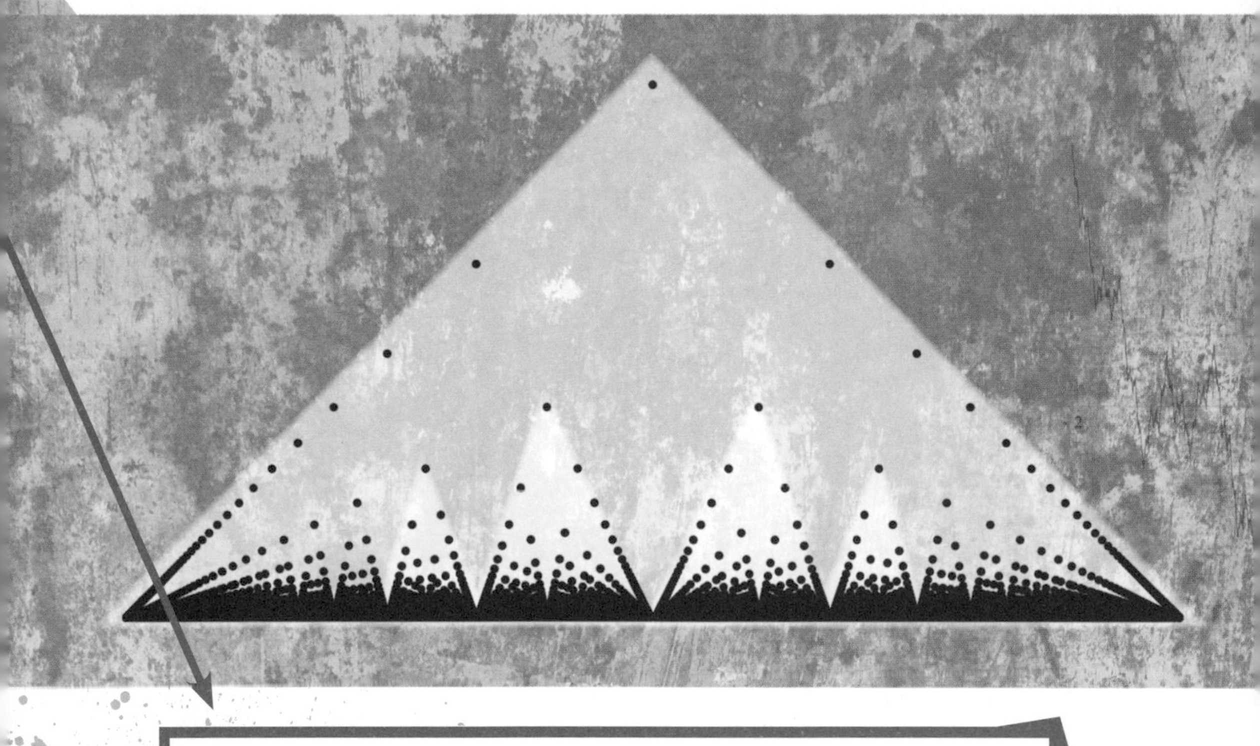

3/ Hack: Pathological functions can warn us that our definitions are too loose, or they can be accepted and open up new avenues of exploration.

If a pathological function gives rise to a single contradiction, it can kill a whole theory.

No.55
Differential Equations
The language of physics

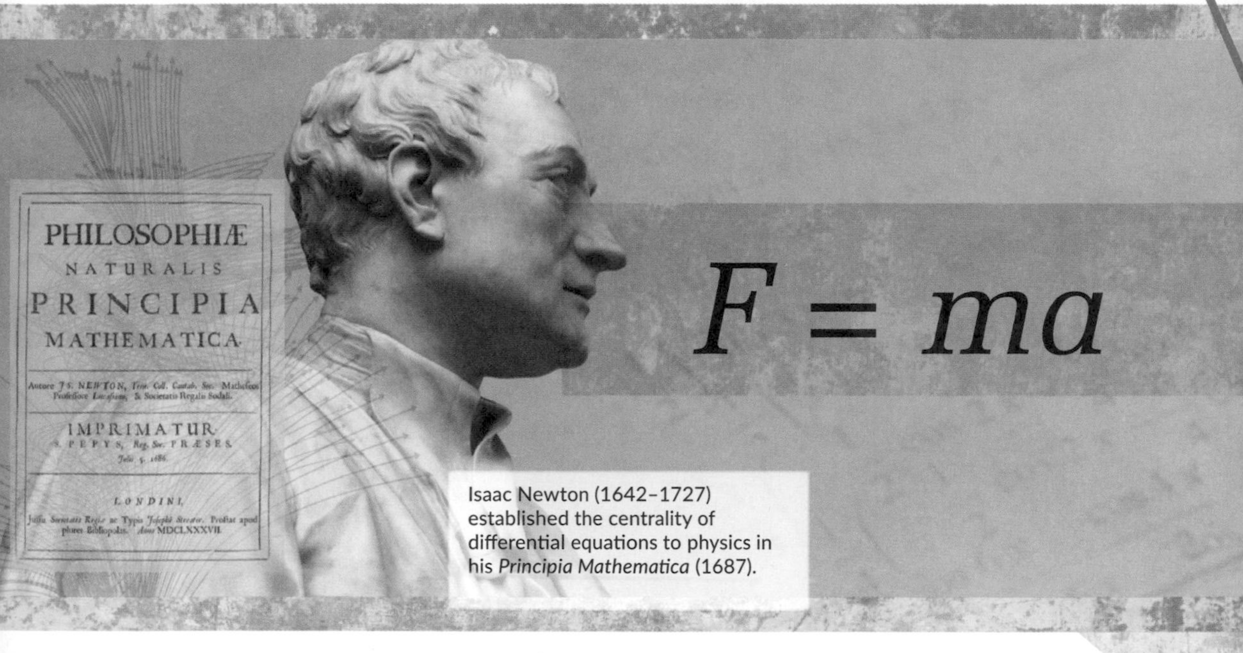

Isaac Newton (1642–1727) established the centrality of differential equations to physics in his *Principia Mathematica* (1687).

1/ Helicopter view: *Differential equations* arise when we know how something changes but not what its actual value is.

For example, Newton's Second Law says $F = ma$, where F is the total force acting on an object, m is its mass, and a is its acceleration. Acceleration is the **derivative** of velocity, which in turn is the derivative of position.

Often we know the forces but want to find out what the object will do, that is, find its position as a function of time. So we rewrite the acceleration as p'', where p is a function mapping time to position. Acceleration is the rate of change of the rate of change of position, so it is the derivative of the derivative of p.

"Solving" this differential equation means finding a function p that makes it true given what we know about F and m. Since they are so ubiquitous, problems like this provided a major motivation for developing and studying calculus.

Right: A slope field is a visual method for finding approximate solutions to differential equations.

2/ Shortcut: The **Fundamental Theorem of Calculus** suggests we can use **integrals** to undo **derivatives**. Unfortunately we do not have a general method that works for all of them, and there are good reasons to suspect we never will. In practical situations only a certain amount of precision is needed, so computer-based approximations are usually used.

For $F = mp''$, things are easy. We can rewrite it as $p'' = F/m$ and integrate twice. Extra information is usually needed to narrow the results down to a unique solution.

See also//

50 The Derivative, p.104

52 The Integral, p.108

53 The Fundamental Theorem of Calculus, p.110

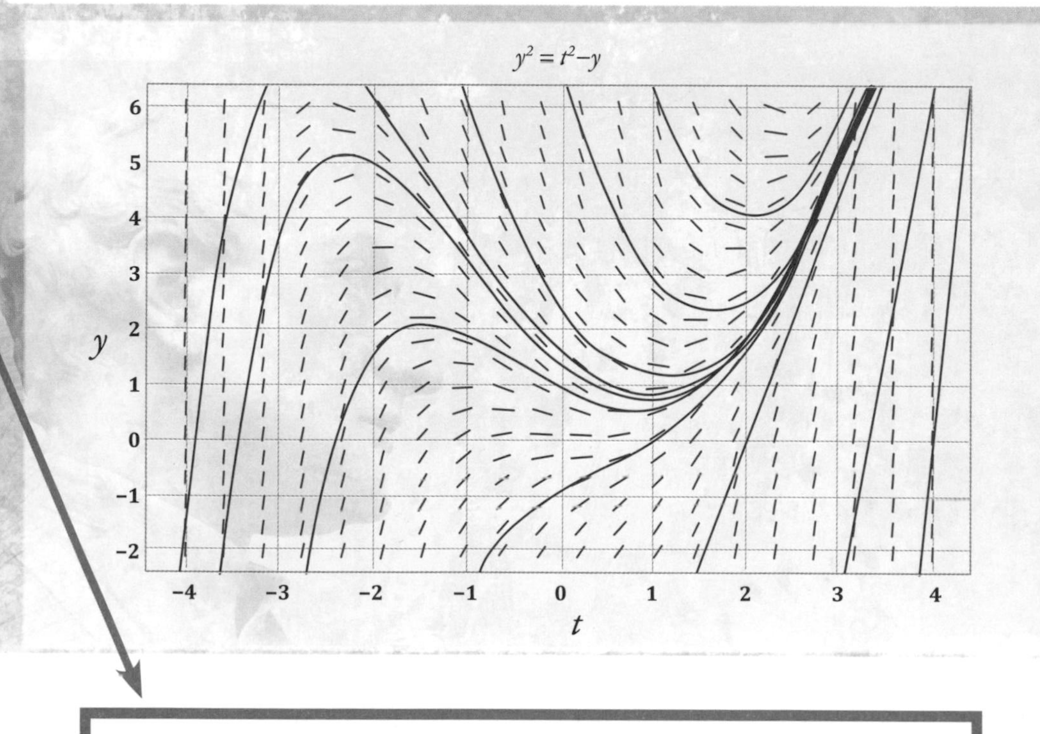

3/ Hack: Differential equations describe change, so they arise in a wide variety of situations. Understanding them better remains a topic of current research.

Logical considerations mean we will probably never have a complete theoretical account of them.

No.56
Calculus of Variations
Optimizing functions

1/ Helicopter view: A *function* **maps** numbers to other numbers; a *functional* maps functions to numbers. You can think of the functional as measuring each function according to some criterion. For example, each function can be graphed, and the graph for values between 0 and 1 is some sort of curvy line. How long is it? A functional can be defined that will tell you.

Often we want to find the function that has the minimum or maximum possible value of a given functional. For example, we might want to find functions whose graphs are as short as possible. This is what the *calculus of variations* is about.

It can be used to prove apparently obvious facts, for example, that a circle encloses the largest area with a fixed length, or that the shortest distance between two points is a straight line. But it generalizes to solve much more difficult problems with only a little extra effort.

Above: Many natural processes, including the behavior of soap films, are well described by the calculus of variations.
Right: When a function hits a local maximum or minimum value, its derivative is zero.

2/ Shortcut:

The main technique in the calculus of variations is to turn a *functional*, which is hard to work with, into a **differential equation**. Though still often tricky, these are much more tractable.

The reason this works is that when a function hits its maximum or minimum value, its **derivative** becomes zero as the thing the functional is measuring "changes direction."

Paths of light rays and the shapes of soap bubbles are among the many physical phenomena that naturally minimize functionals in this way.

See also//
9 Maps, p.22
50 The Derivative, p.104
55 Differential Equations, p.114
66 Minimal Surfaces, p.136

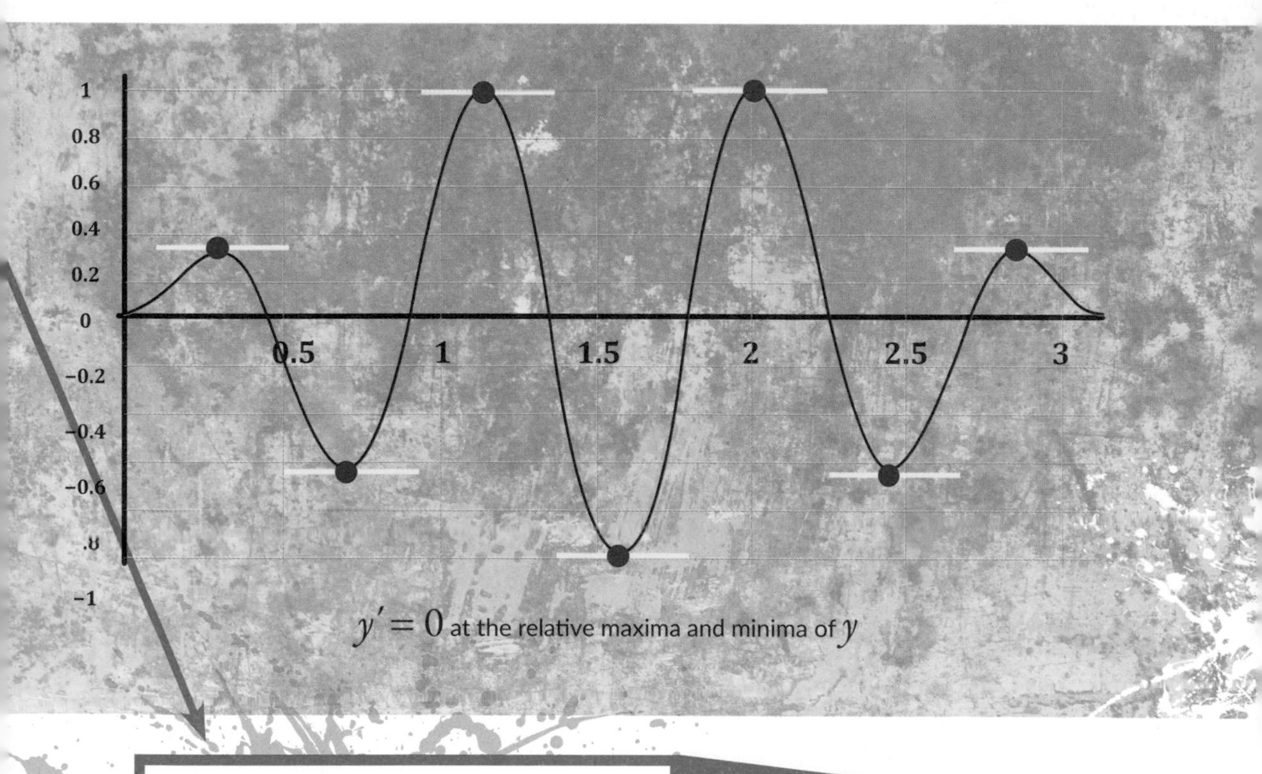

$y' = 0$ at the relative maxima and minima of y

3/ Hack:

Functionals measure functions; techniques from the calculus of variations can help us find maximum or minimum values for these measures.

Often, the trick is to convert the problem into a differential equation that can then (hopefully) be solved.

No.57 Vectors

Numbers with a sense of direction

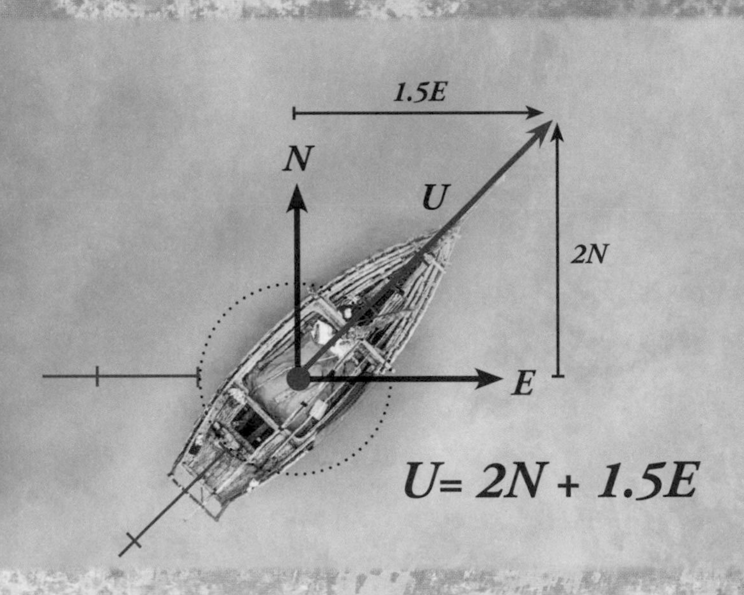

1/ Helicopter view: You can start by picturing a *vector* as a little arrow floating in space. Perhaps it represents the speed and direction of the wind at that point. You could describe such a vector by a list of numbers that indicate how long the arrow is in each direction. In a two-**dimensional** space the vector needs two numbers, say, one indicating how far north it points, the other how far east. Thus, a vector can be thought of as a little arrow representing a displacement in space.

More formally, we define a *vector space*. This consists of *two sets* with some extra structure. One is called the set of *scalars*, which must have the structure of a **field**. The other is the set of *vectors*, which has the structure of a **commutative group**, which we think of as the ability to add two vectors to get another vector.

To complete the vector space we define an operation of "multiplying a vector by a scalar," which we think of as "scaling the vector's size."

Above: The bearing of a boat can be described in terms of its northward and eastward trajectories; the result is a vector. *Right:* Attaching a vector at every point in a space can simulate a flow such as the currents in a river.

2/Shortcut:

A *vector space* is a very abstract object that seems to have nothing to do with arrows or lists of numbers. This enables us to develop a very powerful theory, which is good because once you take the abstract view you find vector spaces everywhere.

Incidentally, a *module* is a vector space except that the scalars form a **ring** rather than a **field**. Modules appear frequently in some areas of mathematics, but their theory is a lot less tidy because rings are much messier than fields.

See also//
35 Abstract Algebra, p.74
38 Groups, p.80
42 Rings and Fields, p.88
64 Dual Vectors, p.132
76 Dimensions, p.156

3/Hack:

Most people picture a vector as a little arrow and calculate with it by thinking of it as a list of numbers.

Officially, a vector is defined as an element of a vector space, which is an abstract algebraic structure.

No.58 Divergence and Curl

Doing derivatives with vectors

1/ Helicopter view: Imagine a 3D space filled with **vectors**—one little arrow attached to every point. This is called a vector field. It might represent the way air is circulating in a room, for example. Can we find a way to describe what is happening close to a given point? For example, is the air rushing away from the point, as if being expelled from a deflating balloon? Or rushing toward it as if someone is taking a deep breath in? The *divergence* of the vector field at that point provides a precise, numerical description of this kind of behavior.

What if the air is circulating around the point, so that a little weathervane placed there would turn in circles? This property of the vector field is measured by its *curl*.

Each operation helps us describe the average or aggregated effect of all the vectors in a tiny region around the point of interest.

Above: Curl occurs when a vector field is "circulating around" at a point. *Right:* Divergence occurs when it is rushing toward or away from the point.

2/ Shortcut: A **vector** field is a **map** from the points of a space to the vectors of some vector space or other. *Divergence and curl* are special **derivatives** that work on this kind of map, and they are a big part of the subject known as *vector calculus*.

They are ultimately defined using the structure of **Euclidean space** itself, which is a product of copies of the real numbers. This means we can reduce things down to maps between numbers, which ordinary calculus works on.

See also//
9 Maps, p.22
50 The Derivative, p.104
57 Vectors, p.118
59 Euclidean Spaces, p.122

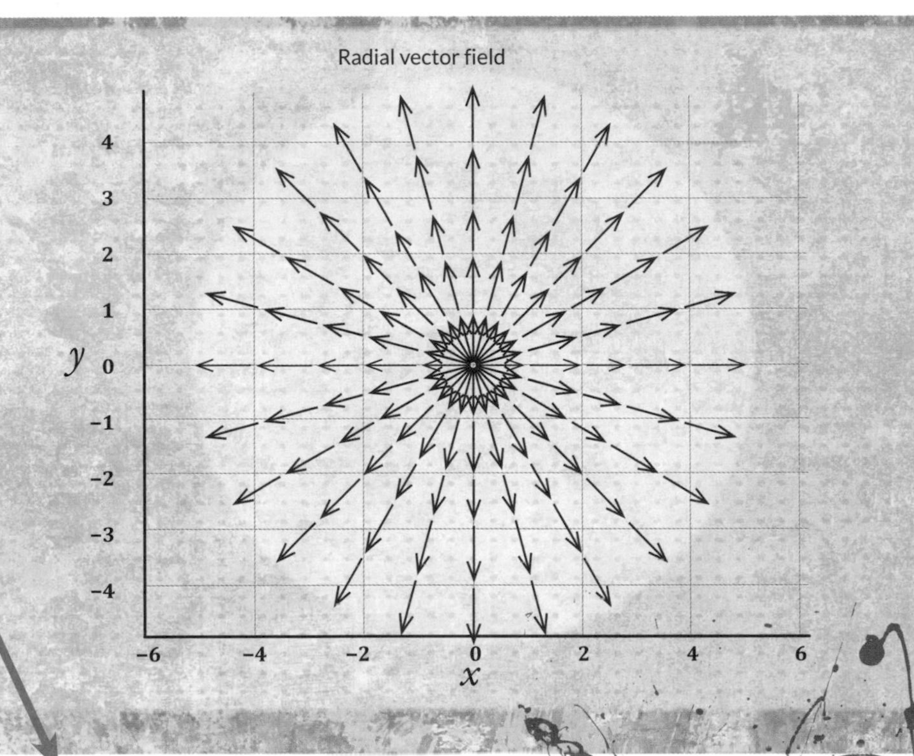

3/ Hack: Divergence measures how much the vectors are flowing toward or away from the point. Curl measures how much they are circulating around it.

These operations describe the behavior of vector fields close to any chosen point.

No.59
Euclidean Spaces
Where ordinary geometry happens

Euclid (c.323–283 B.C.E.)

1/ Helicopter view: The page where school geometry happens is a flat two-**dimensional** (2D) space. This is also where most of the **axioms** and **theorems** from Euclid's book *Elements* take place. The shortest distance between any two points in this space is the straight line you draw between them with a ruler.

We can add a dimension to get 3D space, although now we cannot draw it quite so easily. Here, again, there is a notion of points in the space and a way to calculate the straight-line distance between them.

The **real numbers** are designed to model the points on a line, which by definition is one-dimensional *Euclidean space*. The distance between points a and b is just $a - b$ or $b - a$, whichever gives a positive answer.

To make a 2D Euclidean space we take the product of this line with itself. At every point, attach a copy of the line. The result is like an infinitely dense, woven fabric.

The dimension of a space says how many coordinates it needs. Polar coordinates (above), in which some coordinates are angles, are sometimes useful but rectangular coordinates (right) are more common.

2/Shortcut: This idea can be repeated to produce 3D, 4D, and higher-**dimensional** *Euclidean spaces*. The points always form the set of **vectors** of a vector space with the scalars being the **real numbers**.

We define a special distance function that maps a pair of points to a number in a way that mimics our intuitive sense of the "straight-line distance." This is called the *Euclidean metric* and is part of the definition of each Euclidean space.

See also//

1 Axiom, Theorem, Proof, p.6
26 Real Numbers, p.56
57 Vectors, p.118
60 Manifolds, p.124
76 Dimensions, p.156

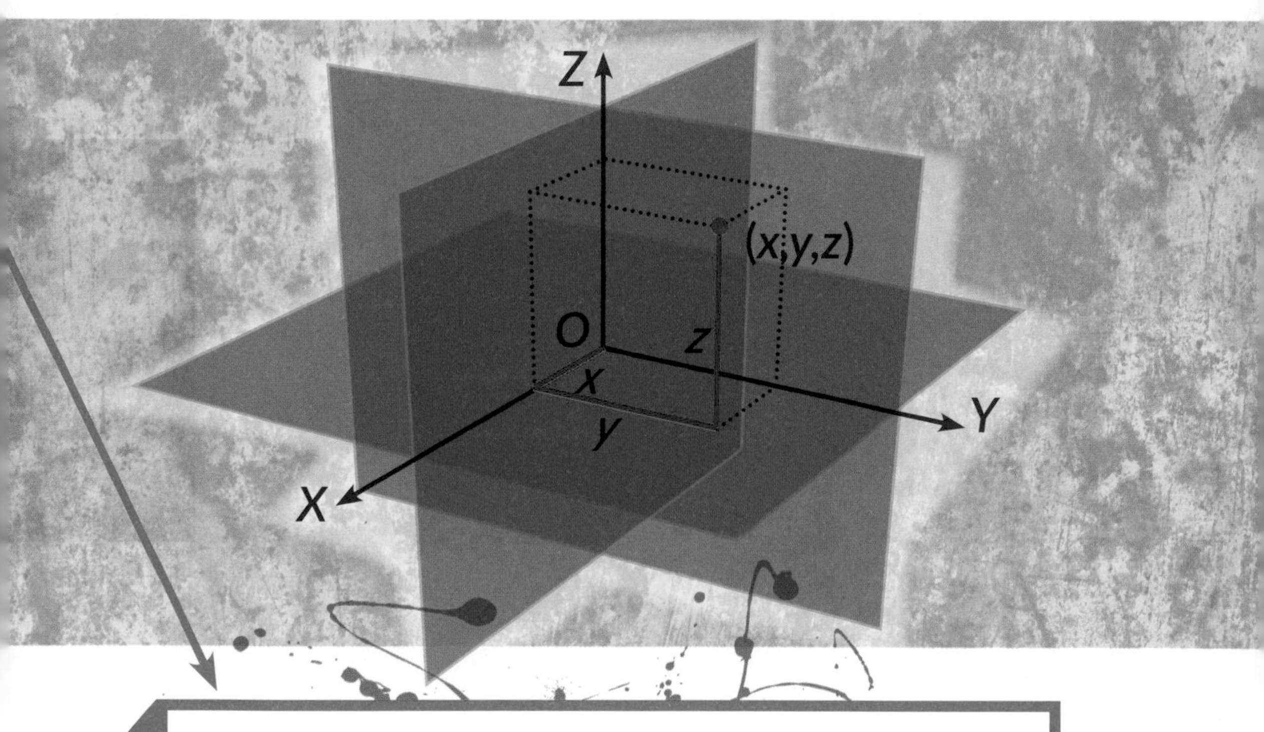

3/Hack: Points in Euclidean space are vectors in a vector space whose scalars are real numbers, plus the intuitively natural way to measure distance between them.

Euclidean spaces are the basis of most other geometries, but they are not the only spaces we can imagine.

No.60

Manifolds

Locally, they look flat

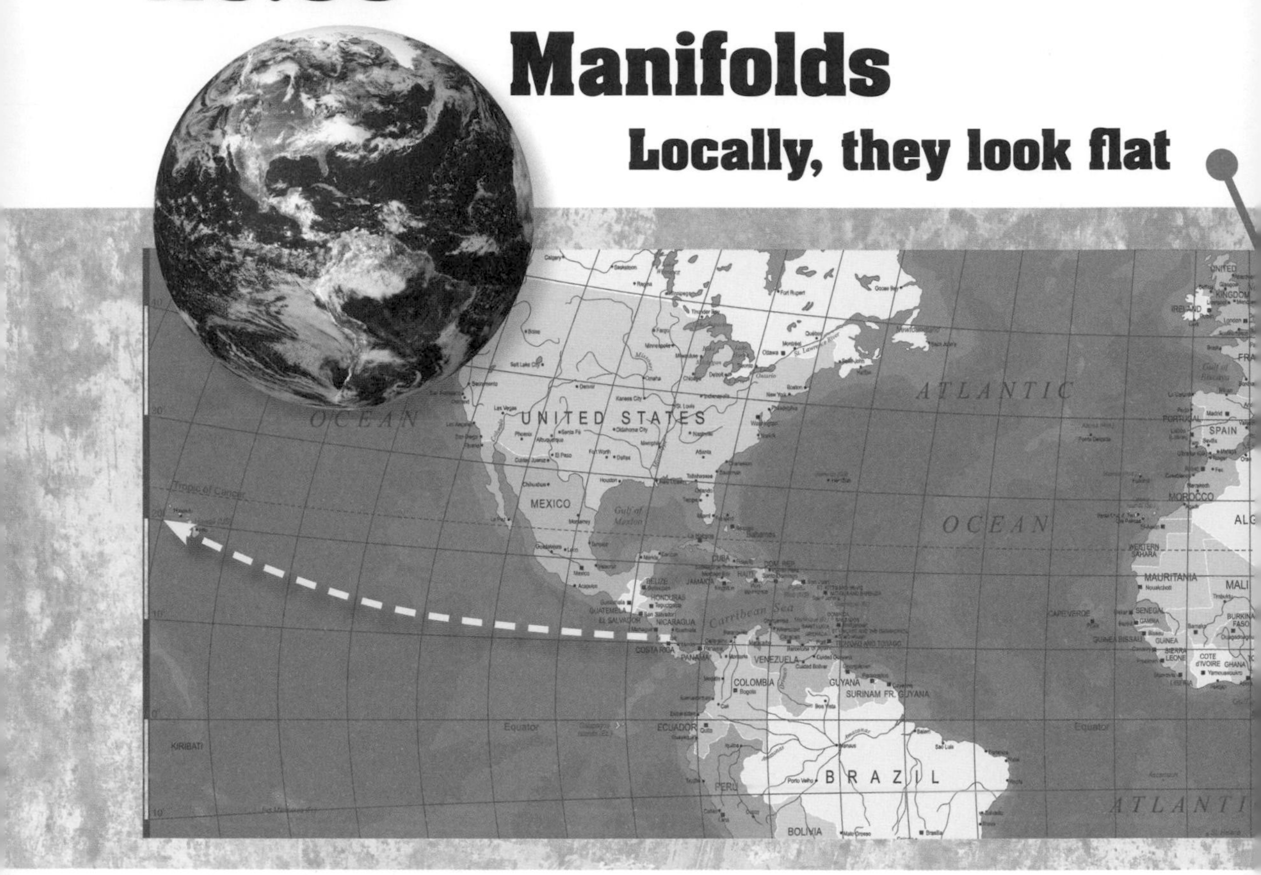

1/Helicopter view: A flat piece of paper is part of two-dimensional **Euclidean space**. What about the surface of the Earth? In a sense it is also two-dimensional because if you stay close to any given point it looks like it might be part of a Euclidean space, allowing for hills and valleys that could be flattened out.

Globally, however, it's nothing like Euclidean space. In fact, we can't make a flat chart (that is, a map, in the nonmathematical sense) of the Earth without "cutting" it somewhere, creating a huge discontinuity where none actually exists. It's not just that the Earth has some incidental curves and wrinkles.

We say the Earth's surface is *locally* Euclidean, meaning that if we pick a small enough region we can make a flat chart of it that only has mild distortions. The fact that we can cover the Earth with overlapping charts like these tells us it's a two-dimensional *manifold*.

Local areas of the Earth can be mapped with a flat chart with little distortion. However, the whole planet can't because it's a manifold that's only locally like a flat Euclidean space.

2/ Shortcut: Making a *manifold* is a matter of finding an atlas of charts that covers the space.

Very roughly, a **topological** manifold is any space such that each point has a region around it that can be cut out and flattened so that it looks like a region of **Euclidean space**.

A *smooth manifold* is one where this can be done without anything sudden happening. A *Riemannian* manifold also includes a way to measure distances in a consistent way.

See also//
59 Euclidean Spaces, p.122
66 Minimal Surfaces, p.136
70 Topology, p.144
75 Curvature, p.154
78 Spherical Geometry, p.160

3/ Hack: Small regions of a manifold look like mildly distorted regions of Euclidean space, but globally they can be connected together in quite different ways.

Spacetime in physics is often modeled as a four-dimensional manifold, with its curvature representing gravity.

No.61 The Tensor Product

New vector spaces from old

Hassler Whitney (1907–89) was probably the inventor of the tensor product.

1/ Helicopter view: Suppose we have two **vector** spaces that share the same scalars and we want to combine them. One way is to form the usual **product** of their **sets** of vectors, then carefully define addition and scalar multiplication to make the end result a vector space.

This does not always give exactly the results we want. For example, to find the **dimension** of the product $V \times W$ we add the dimensions of V and W. But if it is a product, should we not *multiply* them instead?

The *tensor product*, $V \otimes W$, is often what we are looking for. This is always another vector space, but in this context its vectors are called tensors. We multiply the dimensions of V and W to get the dimension of $V \otimes W$.

The tensor product solves a special algebraic problem. Roughly, every "reasonably nice" (technically, "bilinear") map from $V \times W$ to another vector space "factors through" $V \otimes W$ in a unique way by a linear map.

Right: The tensor product can turn a map that isn't quite linear (h) into a linear one ($\tilde{h}$).

2/Shortcut: Take two **vector** spaces V and W that share the same scalars. Their *tensor product* $V \otimes W$ is a new vector space with the same scalars. Its vectors are all the possible pairings of an element of V with an element of W, along with all the ways we could add several of those together.

For example, a vector in the new space might look like $v_1 \otimes v_1 + v_2 \otimes w_2$, where v_1 and v_2 are vectors in V and w_1 and w_1 are vectors in W.

See also//
7 Set Theory, p.18
8 Products, p.20
57 Vectors, p.118
76 Dimensions, p.156

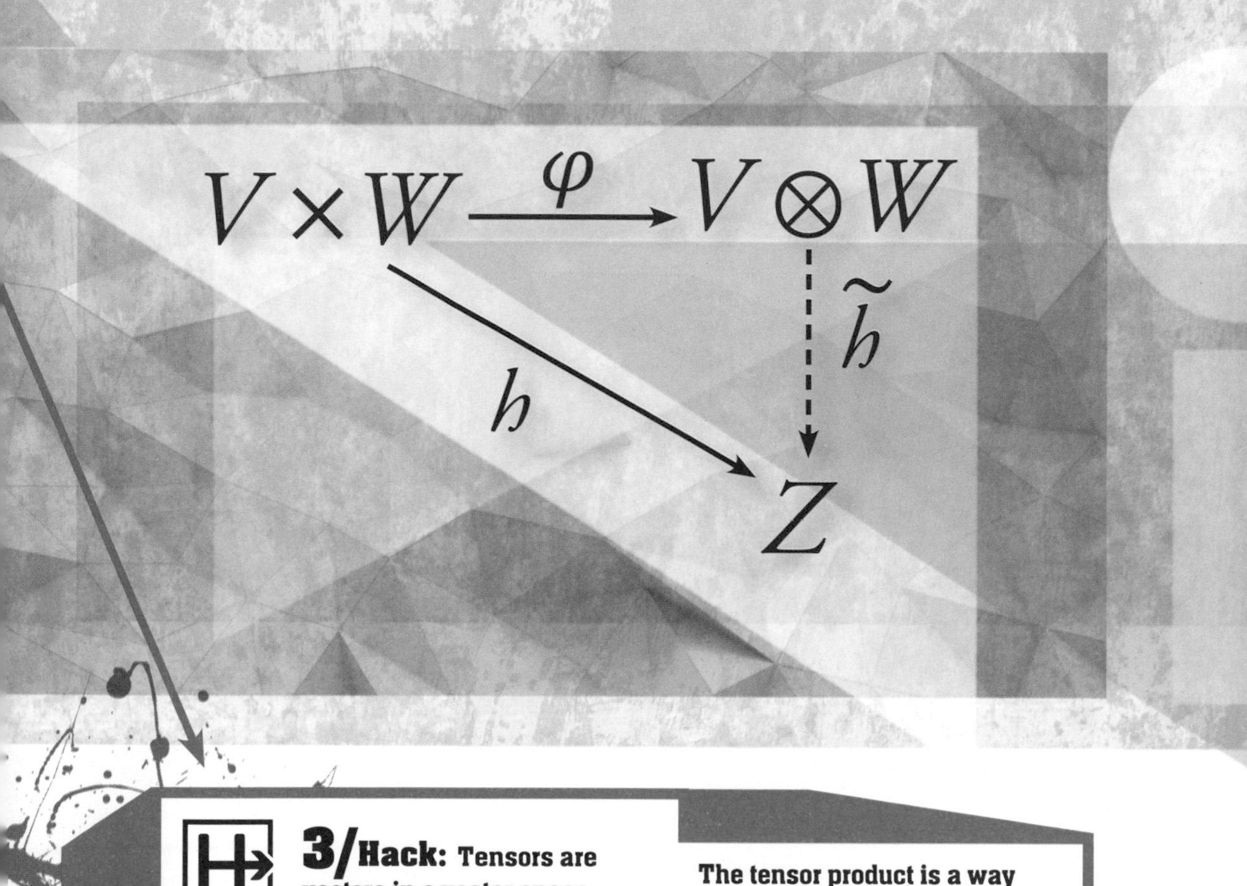

3/Hack: **Tensors are vectors in a vector space that was made from other vector spaces using the tensor product.**

The tensor product is a way to multiply vector spaces together to give a new vector space.

No.62 Covariant and Contravariant

They look similar but transform differently

1/ Helicopter view: Suppose we are looking at a moving particle, measuring distance in kilometers and time in seconds. We observe that the particle is traveling at 1km/s. Now change the units of distance to meters instead. The point, which has not changed its physical behavior, is now traveling at 1,000m/s.

This change from 1 to 1,000 is a scaling of the velocity **vector** that happens because of we transformed the way we are measuring position. Note that nothing physical has changed, just the number we use to describe it.

Now suppose that, during a journey of 1km, the particle experiences a rise of temperature of 1°C. We say the *temperature gradient* was 1°C per km. Now apply the transformation from km to meters again. This time the temperature gradient is 0.001°C per meter.

But this time the number has *shrunk* by a factor of 1,000, whereas before it *grew*. What is different this time?

Below: When we reduce the unit of horizontal measurement from km to m, the number measuring the skiers' velocities contravaries but the measure of the slope covaries. *Right:* Changing coordinate systems is very common; we must always be aware of how this affects other measurements.

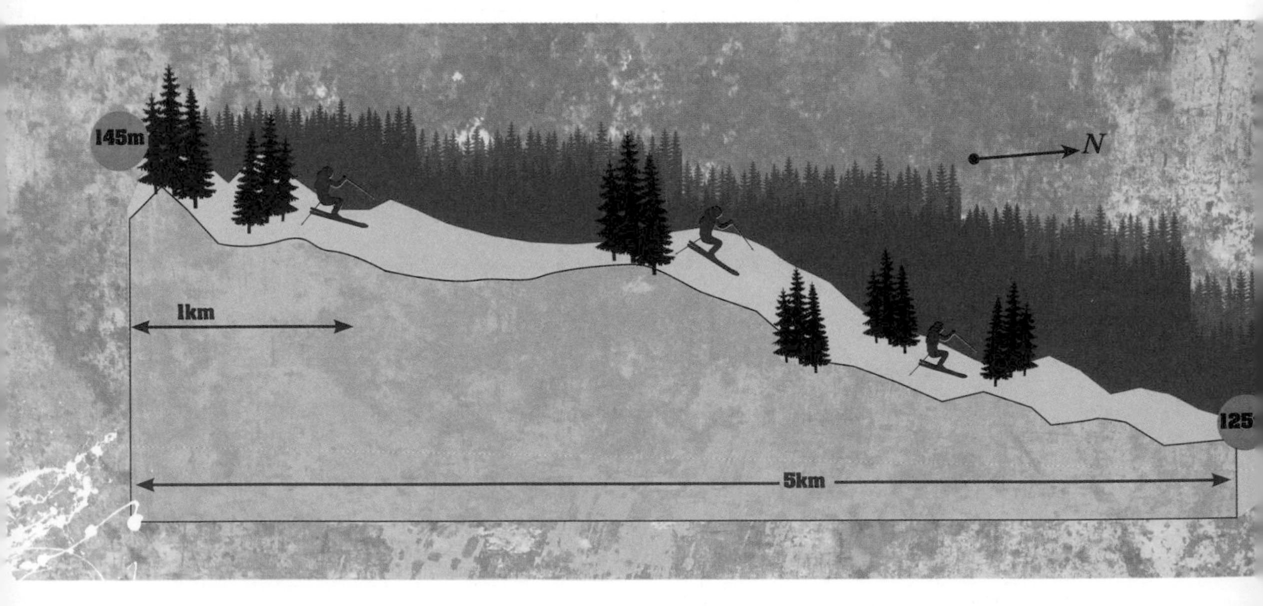

2/ Shortcut: We say the velocity **vector** was *contravariant* because we made the units of measurement 1,000 times *smaller* but the vector's size seemed to get 1,000 times *bigger*.

The gradient, on the other hand, got 1,000 times smaller—the same as the units of measurement—so we call it *covariant*.

Nothing is physically happening, just the frame of reference with which we are looking at the situation. But switching frames of reference is a key part of physics, so these distinctions are important.

See also//

57 Vectors, p.118

64 Dual Vectors, p.132

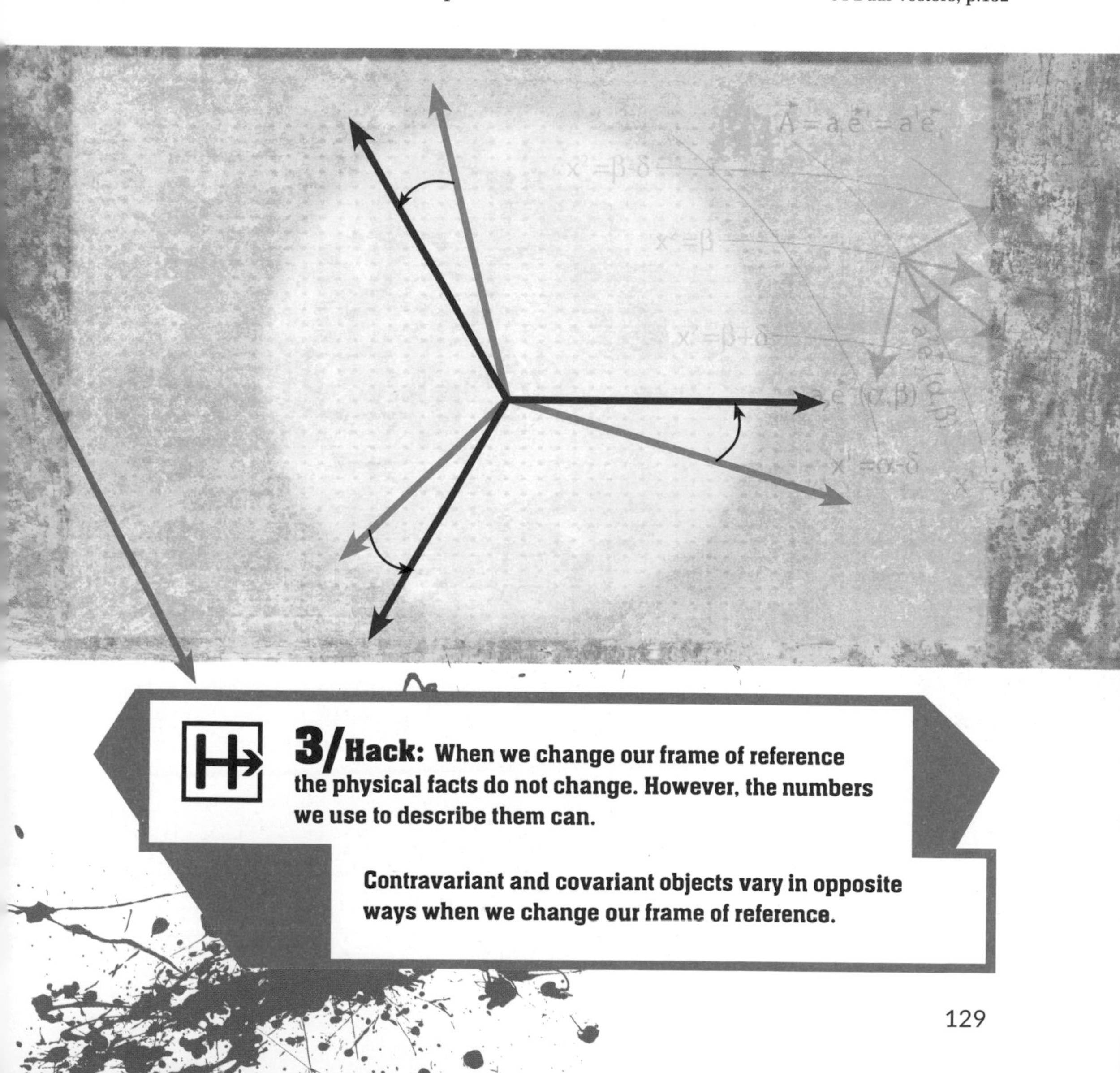

3/ Hack: When we change our frame of reference the physical facts do not change. However, the numbers we use to describe them can.

Contravariant and covariant objects vary in opposite ways when we change our frame of reference.

No.63 Matrices Nice transformations of vector spaces

1/ Helicopter view: Any **map** from a **vector** space to itself can be called a transformation, but most maps you could define in this way are quite nasty and have limited theoretical or practical use.

But there are a few families of transformations that are especially important. Most important of all are the *linear transformations*, which preserve the algebraic structures of the two spaces. These are sometimes called the *automorphisms* of *V*, and they form a very important **group**.

In elementary mathematics the linear transformations from *V* to itself are represented as squares of numbers with some algebraic rules attached; such a square is called a *matrix*. Which matrix represents the transformation, however, depends on choosing a basis for each vector space.

You can also define linear transformations between two different vector spaces. In this case sometimes the matrix is rectangular rather than square. The study of vector spaces and linear transformations is called *linear algebra* and is extremely well developed.

Vector fields can measure physical phenomena from pressure gradients to electromagnetism; their elements might be contravariant (vectors) or covariant (dual vectors).

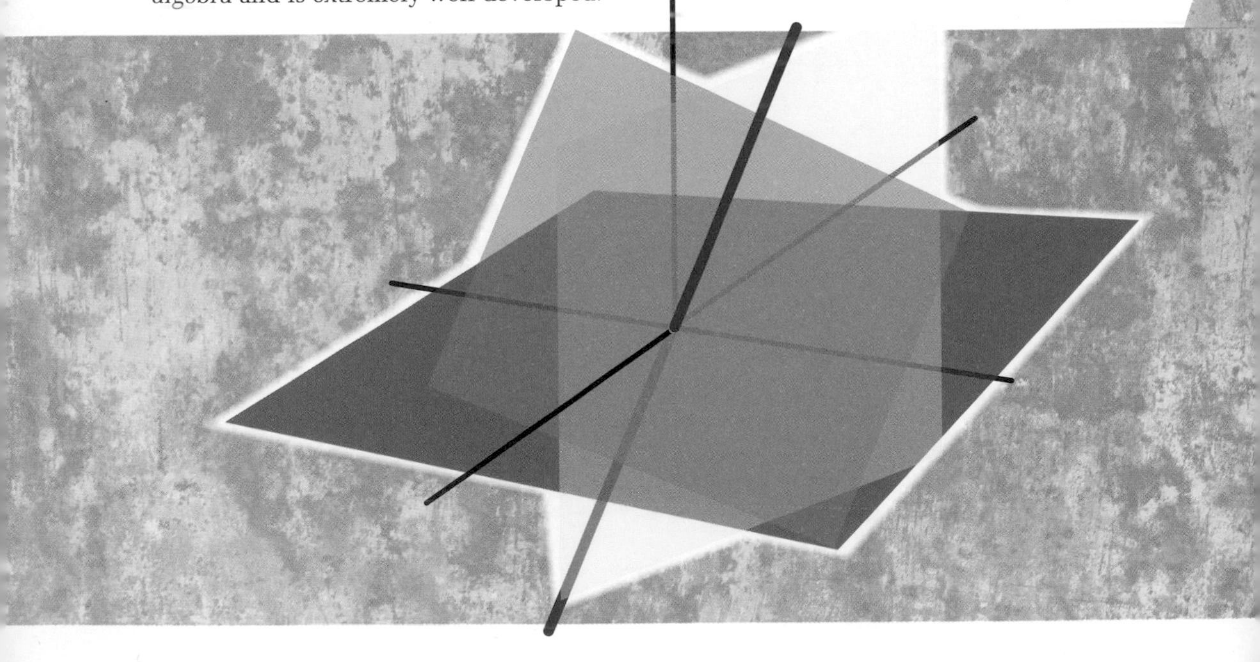

2/ Shortcut: For a map between **vector** spaces to be linear we should have, for vectors v and w and a scalar s:

$$f(v+w) = fv + fw$$
$$f(sv) = sf(v)$$

Instead of a matrix, you can think of an automorphism of V as belonging to the **tensor product** $V \otimes V^*$. Very roughly, the covector on the right acts on any vector in V to produce a scalar, which then multiplies by the vector on the left to give the result.

See also//

35 Abstract Algebra, p.74
38 Groups, p.80
57 Vectors, p.118
61 The Tensor Product, p.126
64 Dual Vectors, p.132

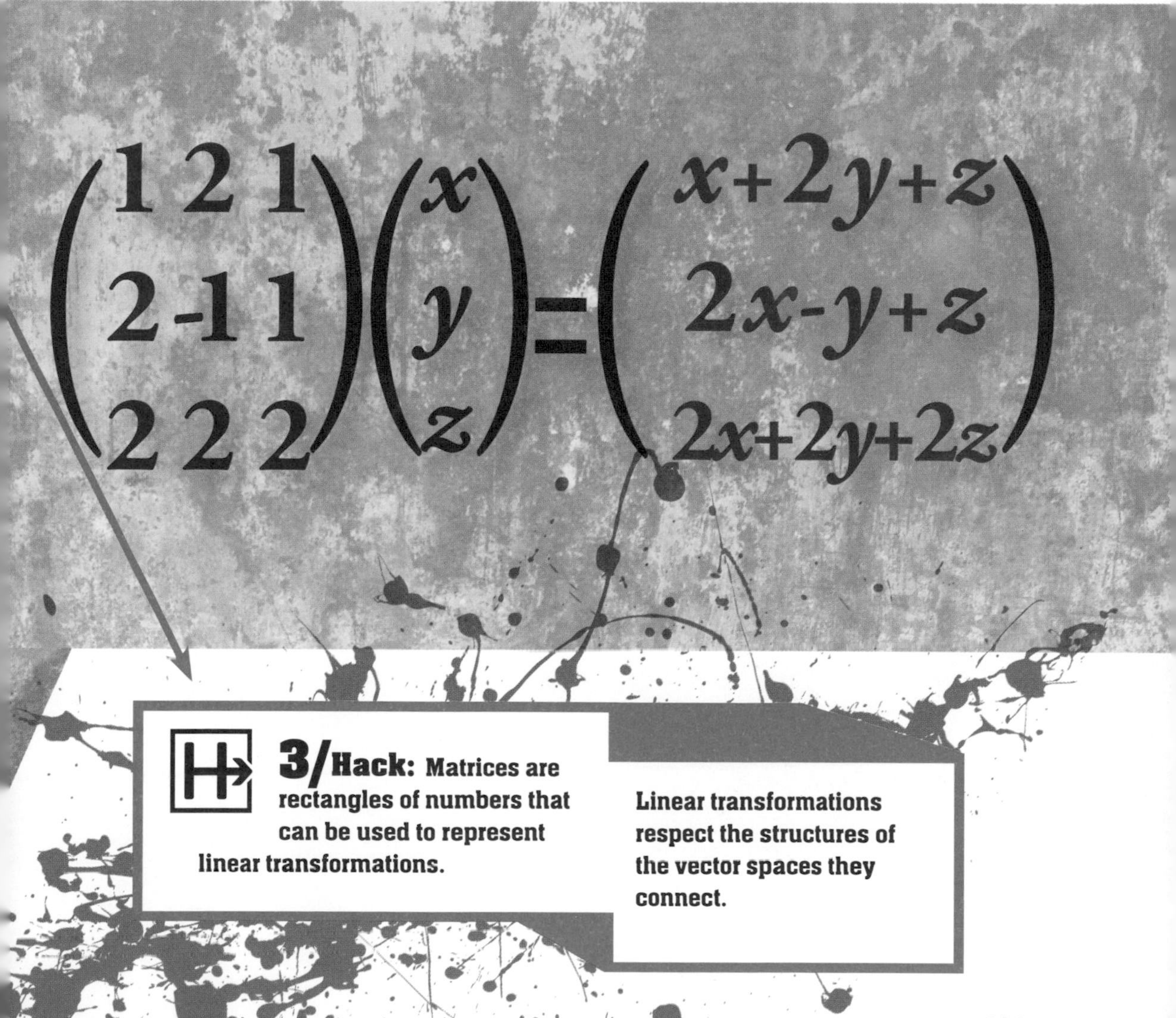

3/ Hack: Matrices are rectangles of numbers that can be used to represent linear transformations.

Linear transformations respect the structures of the vector spaces they connect.

No.64 Dual Vectors

Every vector space has its mirror image

1/ Helicopter view: Suppose we have a **vector** space V. A *functional* is a **map** from the vectors in V to the scalars. There are many kinds of functional, but some obey certain rules that make them linear.

It turns out that the set of all linear functionals that can be defined on V is itself a vector space with the same set of scalars, written V^*. In fact, in an **abstract algebraic** sense V^* is isomorphic to V and we can map one to the other precisely without losing any structure. The linear functionals are then called *dual vectors* or *covectors* and V^* is called the *dual space*.

If we form the dual space of the dual, V^{**}, we get back to V again. So V and V^* really are like mirror reflections of each other.

Although vectors and covectors intuitively look like little arrows, the way they transform means each is suitable for modeling different physical facts.

A linear transformation of space can be represented by a matrix.

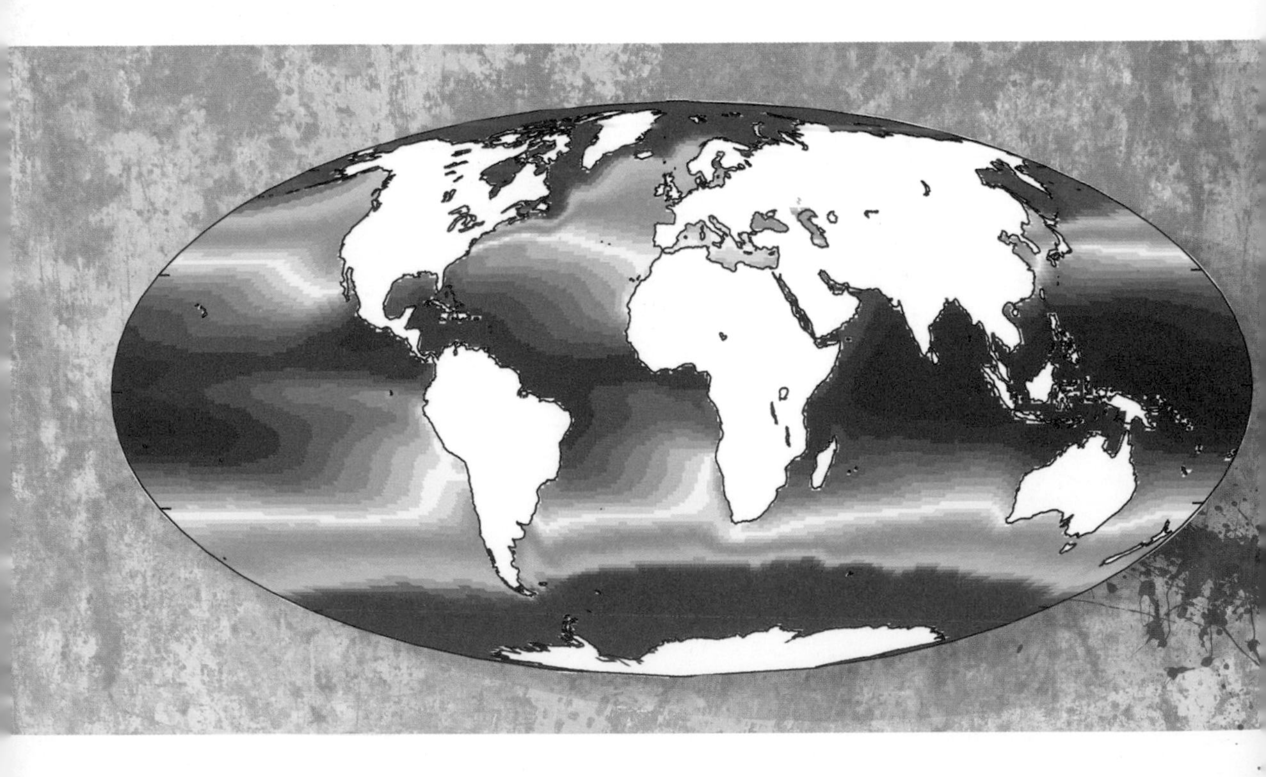

2/ Shortcut:

A functional f is linear if it obeys the following laws, where v and w are vectors and s is a scalar:

- $f(v + w) = f(v) + f(w)$
- $f(s \times v) = s \times f(v)$

The difference between V and V^* is revealed when we change the *basis* of V (or V^*), which you can think of as a frame of reference for describing vectors precisely.

The vectors in V transform **contravariantly** just as we expect a measure of velocity would. But the covectors in V^* transform **covariantly**, like a measure of gradient.

See also//
9 Maps, p.22
35 Abstract Algebra, p.74
55 Differential Equations, p.114
57 Vectors, p.118
62 Covariant and Contravariant, p.128

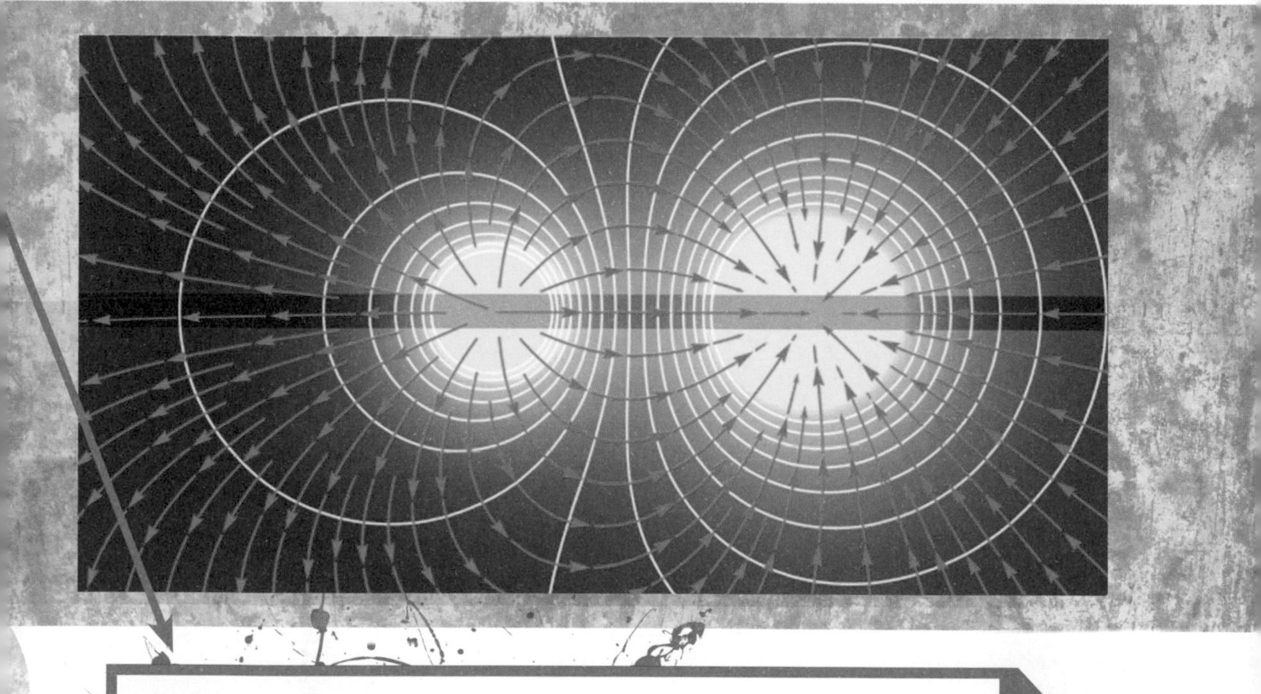

3/ Hack:

The dual of a vector space is like its mirror image, with the same scalars and linear functionals as its vectors.

Every vector space has its dual; if one contravaries with a transformation, the other covaries and vice versa.

No.65
Tensor Fields
Things that happen in space

1/ Helicopter view: Physical situations are often modeled as an empty space with something attached to each point. The most obvious thing to attach to each point is a number. It might represent the temperature there, for example. This is called a *scalar field*.

We can attach a **vector** to each point instead. If we imagine it as a little arrow, we can think of it as representing some kind of movement in the space at that point, such as air flow. This is either a **vector** field or *covector field* (more usually called a 1-form).

Some physical phenomena are more complicated, though. They can sometimes be captured by attaching a tensor at each point. It might come from taking the **tensor product** of a vector space V with itself multiple times, and perhaps with its dual space V^* as well.

This is a *tensor field*. A tensor squeezes a lot of information into a form that is well behaved and easy to work with.

Experimental visualizations of tensor fields devised by Ingrid Hotz, Louis Feng, Hans Hagen, Bernd Hamann, Boris Jeremic, and Kenneth I Joy.

2/ Shortcut: Think of a tensor as a "package" containing a mixture of several vectors and covectors. When we change our frame of reference, they all automatically transform as they should, which makes them useful for doing physics.

In a curved space such as a smooth **manifold**, even such basic geometric concepts as length and angle are underwritten by a *tensor field* (a *metric* that attaches a little machine for calculating them to every point). It varies from point to point depending on how the space curves.

See also//

55 Differential Equations, p.114

57 Vectors, p.118

60 Manifolds, p.124

61 The Tensor Product, p.126

62 Covariant and Contravariant, p.128

3/Hack: Complicated physical phenomena spread across a region of space can often be modeled by tensor fields.

Attach to each point a little package of vectors and covectors from a tensor product.

No.66 Minimal Surfaces

Can soap bubbles do calculus?

1/Helicopter view: Mix up some soapy water and dip a wire frame of any shape into it. The film will cling to the frame in a particular way. Often the shape it makes does not look obvious to us, but it is always the same for a given piece of wire.

This shape is the result of the molecules of the soapy water wanting to cling together. They do not have a global plan; each molecule only sees the ones close to it, but it tries to pull them all toward itself. The result is a surface that, under the constraints imposed by the wire frame, minimizes its surface area in the small region surrounding each molecule. In other words, it solves a **calculus of variations** problem.

Since nature often seems to follow principles of "least action," the tools we use to investigate these objects have many applications beyond the obvious geometric ones. As shapes, they are also widely used in architecture and design.

The shapes formed by soap films can be described as solutions to problems in the calculus of variations.

2/ Shortcut: A *minimal surface* is one in which the following is true; we can find a region around any point on the surface whose area is the smallest it can be given the shape and length of the region's boundary.

All the interesting examples are curved, which means measurements of length and area can vary from point to point. This means a *metric* **tensor field** is needed before they can be rigorously defined.

See also//

56 Calculus of Variations, p.116
60 Manifolds, p.124
65 Tensor Fields, p.134
74 Metric Spaces, p.152

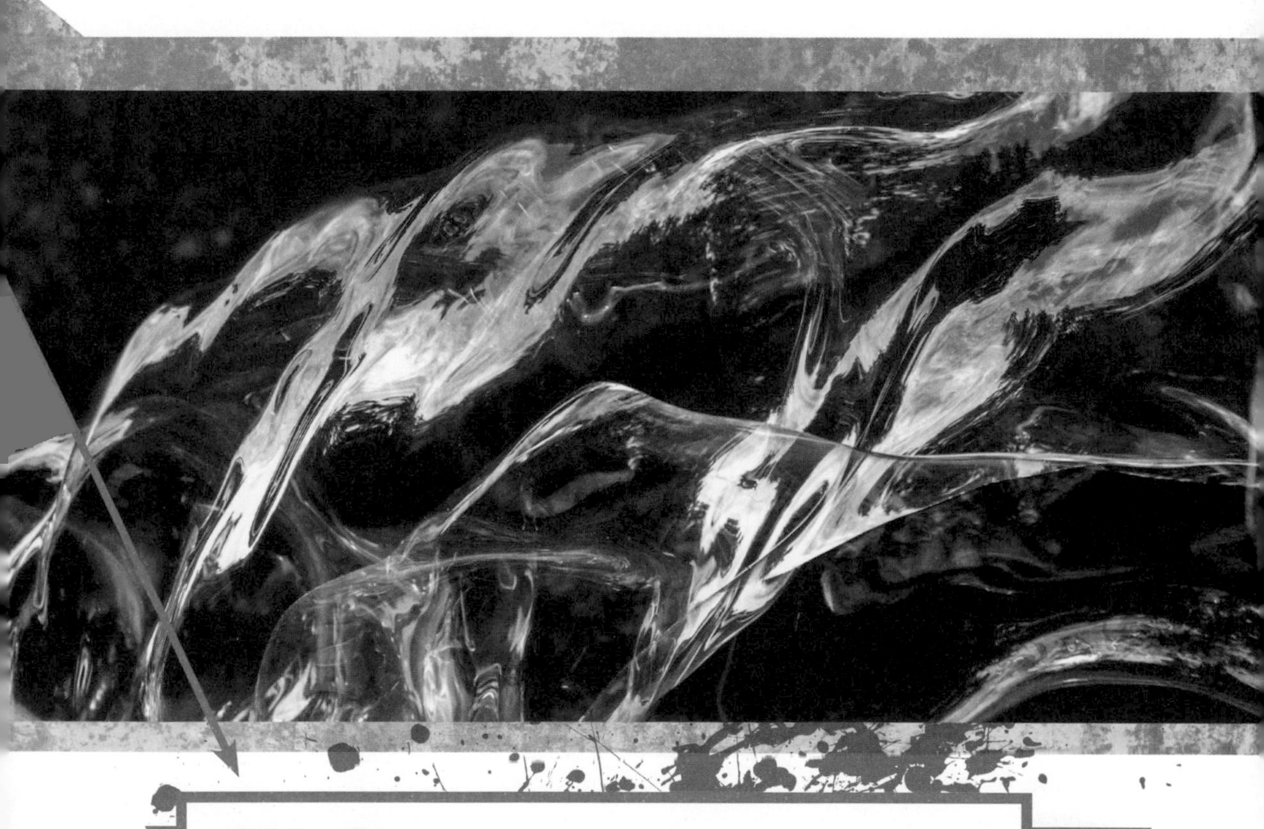

3/ Hack: A minimal surface minimizes its surface area locally, relative to the way of measuring areas and lengths that makes sense there.

Each point can only "see" points close to it, so area is only minimized in small regions.

No.67 Representation Theory

Making models with matrices

1/ Helicopter view: A **group** is a purely abstract object. Its elements can represent many different kinds of concrete things, such as numbers, symmetries of a crystal, electron orbits, field automorphisms (in **Galois theory**), elements of a cryptographic system, moves in a game, continuous transformations of a space, and even just everyday objects. Group theory usually hides all that complexity, making the structure easier to see.

One context where groups naturally emerge is in studying collections of square **matrices**, which can be thought of as geometric transformations of a **vector** space. Multiplying two such matrices has the effect of applying one transformation followed by the other. Under the right conditions, a collection of matrices can form a group with multiplication as its **binary operation**.

The surprising thing is that this works the other way around, and any group—regardless of how we discovered it—can be represented by a collection of matrices. This is the perspective of *representation theory*.

Matrices—squares of numbers governed by algebraic rules—can give a concrete representation of the symmetries of complicated objects.

2/Shortcut: **Groups** arise because we separate the structure from some concrete mathematical objects in order to study it, making them very abstract. Representations enable us to make groups concrete again, but in a consistent way.

This turns every group into a collection of **matrices**—which we already know a lot about—so it is useful for people who study group theory. It can shine new light on matrices, too, meaning that this theory unifies two of the most common objects in all of mathematics.

See also//
36 Binary Operations, p.76
38 Groups, p.80
43 Galois Theory, p.90
63 Matrices, p.130

To rotate counterclockwise by:

90 degrees	180 degrees	270 degrees
multiply by:		
$\begin{bmatrix} 0 & -1 \\ 1 & 0 \end{bmatrix}$	$\begin{bmatrix} -1 & 0 \\ 0 & -1 \end{bmatrix}$	$\begin{bmatrix} 0 & 1 \\ -1 & 0 \end{bmatrix}$

3/Hack: Group theory comes from abstracting a structure from a wide range of concrete situations. Representation theory gives us a consistent way to make the structure concrete again.

Connecting groups with matrices in this way brings seemingly unrelated topics in mathematics closer together.

No.68 Parallel Lines

Never the twain shall meet

1/ Helicopter view: In a nutshell, two lines are *parallel* if they never meet (the technical term is "intersect").

Parallel lines are an essential part of Euclid's geometry. Here they are always straight lines that *go on forever in both directions.* Euclid's Fifth Postulate says that if you give me one straight line and any other point in two-dimensional space, I can draw one and only one line through the point that is parallel to the line you gave me.

This is one of Euclid's **axioms** but for millennia mathematicians tried to prove it from the others, so that it would not have to be assumed. After all, if you play with it for a while it looks like simply a fact about how lines and points work. It turns out, though, that it's impossible to prove it. This impossibility led to the development of the non-Euclidean (**spherical** and **hyperbolic**) **geometries** in the 19th century.

Above: Parallel lines never end and never meet. The contour lines on a map are parallel curves. *Right:* Parallel straight lines may appear to meet, depending on how you look at them.

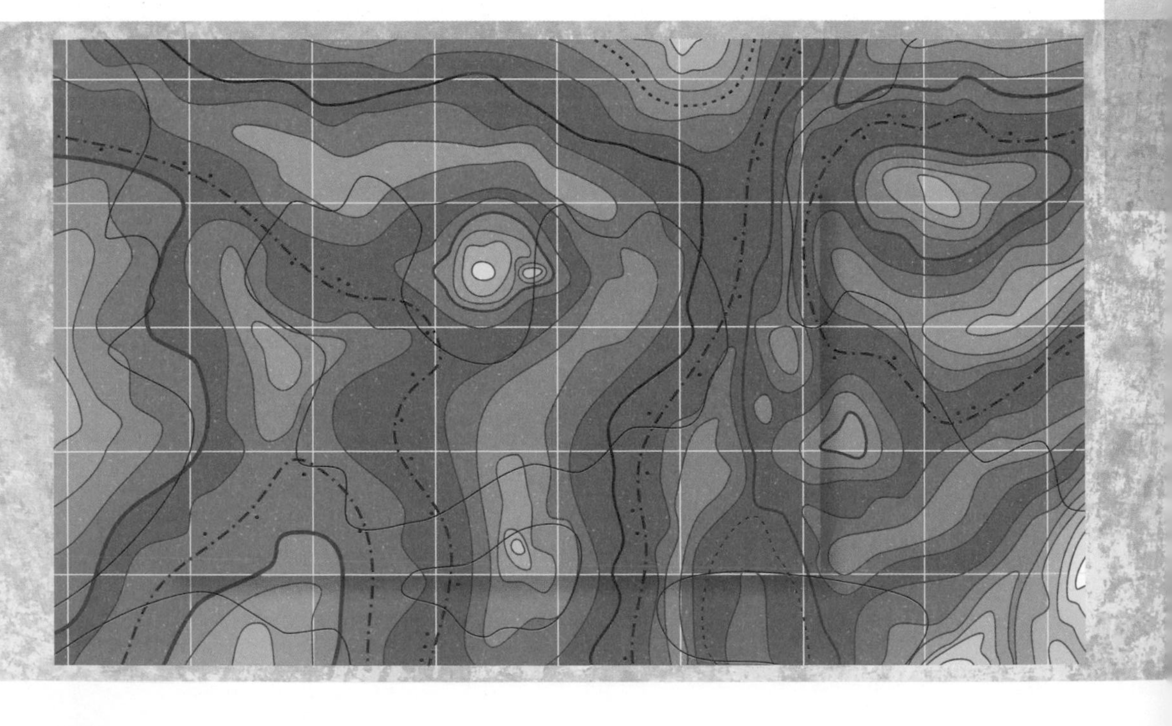

2/Shortcut: Straight lines that do not go on forever have end-points; we usually call them *segments*. Infinitely long straight lines, of course, do not have end-points. Euclid's Fifth Postulate only applies to infinite straight lines.

There is another kind of line like this, which is a loop whose ends are connected. Examples of this include circles, ellipses, and all deformed versions of these. Two loops like this are sometimes said to be *parallel* if they never meet, that is, if one lies entirely inside the other, like the contour lines of mountains on a map.

See also//
1 Axiom, Theorem, Proof, p.6
59 Euclidean Spaces, p.122
78 Spherical Geometry, p.160
79 Hyperbolic Geometry, p.162

3/Hack: The uniqueness and existence of parallel straight lines is fundamental to 2D Euclidean geometry. It cannot be proved from its other axioms.

Two lines are parallel if they do not have end-points and never meet.

No.69 Impossible Constructions

All theories have their limits

1/ Helicopter view: **Euclidean** geometry begins with a small collection of **axioms** and goes on to show how to use them to solve a dazzling array of problems. One such is this. Suppose I give you two lines that meet at an angle. Can you cut that angle in half? The answer is yes, and Euclid's methods can do it very easily.

However, other problems are more difficult. Suppose I give you an angle and ask you to cut it into thirds? This is called the problem of *trisecting the angle*.

Another is this. Suppose I give you a circle. Can you draw a square whose area is the same as the circle's? This is the problem of *squaring the circle*.

Finally, a 3D example; suppose I give you a cube. Can you give me back a cube of twice the volume? This is, unsurprisingly, called *doubling the cube*.

For centuries people tried to find solutions to these ancient puzzles. We now know they are all *impossible*.

Right: An attempt to "square the circle" using Euclidean methods. As with other methods, it only works approximately.

2/ Shortcut: What makes these problems seductive is that they look very much like other problems that are easy to solve within Euclid's system. It's not at all obvious that they should be difficult, never mind *impossible*.

They can all be solved, but only by adding to Euclid's **axioms**, making his geometry much more powerful. This fact was only properly understood in the 19th century, when it was deeply connected with the development of **Galois theory** and the rise of **abstract algebra**.

See also//

1 Axiom, Theorem, Proof, p.6
35 Abstract Algebra, p.74
43 Galois Theory, p.90
46 The Unsolvable Quintic, p.96
59 Euclidean Spaces, p.122

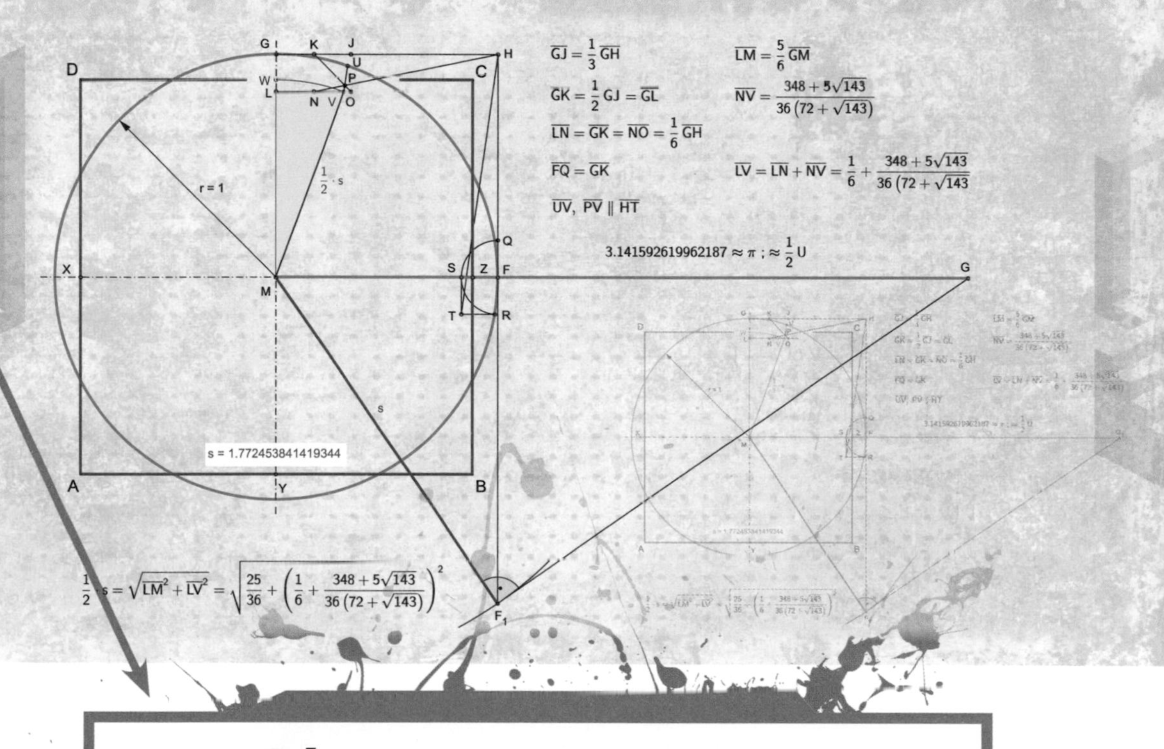

3/ Hack: The axioms of any theory allow us to prove things, but no theory can prove everything. These are examples of the limits of Euclid's geometry.

Trisecting angles, squaring circles, and doubling cubes are all impossible if we restrict ourselves to Euclid's axioms.

No.70 Topology

Geometry without measurement

1/ Helicopter view: The word "geometry" probably comes from its earliest applications, such as practical problems in surveying and architecture. At the heart of such problems are questions about lengths, angles, areas, and volumes (things that can be *measured*).

Not everything interesting about space can be measured in this way. Consider the torus, which is a fancy name for the shape of a doughnut with a hole. Compare it to a sphere. The most striking difference between them is that one has a hole and the other does not.

Geometric methods can tell us things about the hole in the torus, but only in over-precise and very cumbersome ways. Perhaps all we want to say is whether the hole is there or not.

This is where *topology* comes in. By stripping away the details of measurement, it concentrates our attention on other aspects of space that geometry tends to obscure.

The torus and Klein bottle are examples of 2D surfaces that have different topological properties.

2/Shortcut:

Since it ignores things like lengths, angles, and so on, *topology* is sometimes called "rubber sheet geometry." To a topologist, any two triangles are identical, because we cannot distinguish between them using their sides or angles. In fact, a triangle, a square, and a circle are all the same, since curves and corners disappear too.

The development of topology was a major achievement of 20th-century mathematics. Although it began as the purest of pure mathematics, it now has many real-life applications.

See also//
72 The Euler Characteristic, p.148
84 Algebraic Topology, p.172
85 Knot Theory, p.174
86 The Poincaré Conjecture, p.176

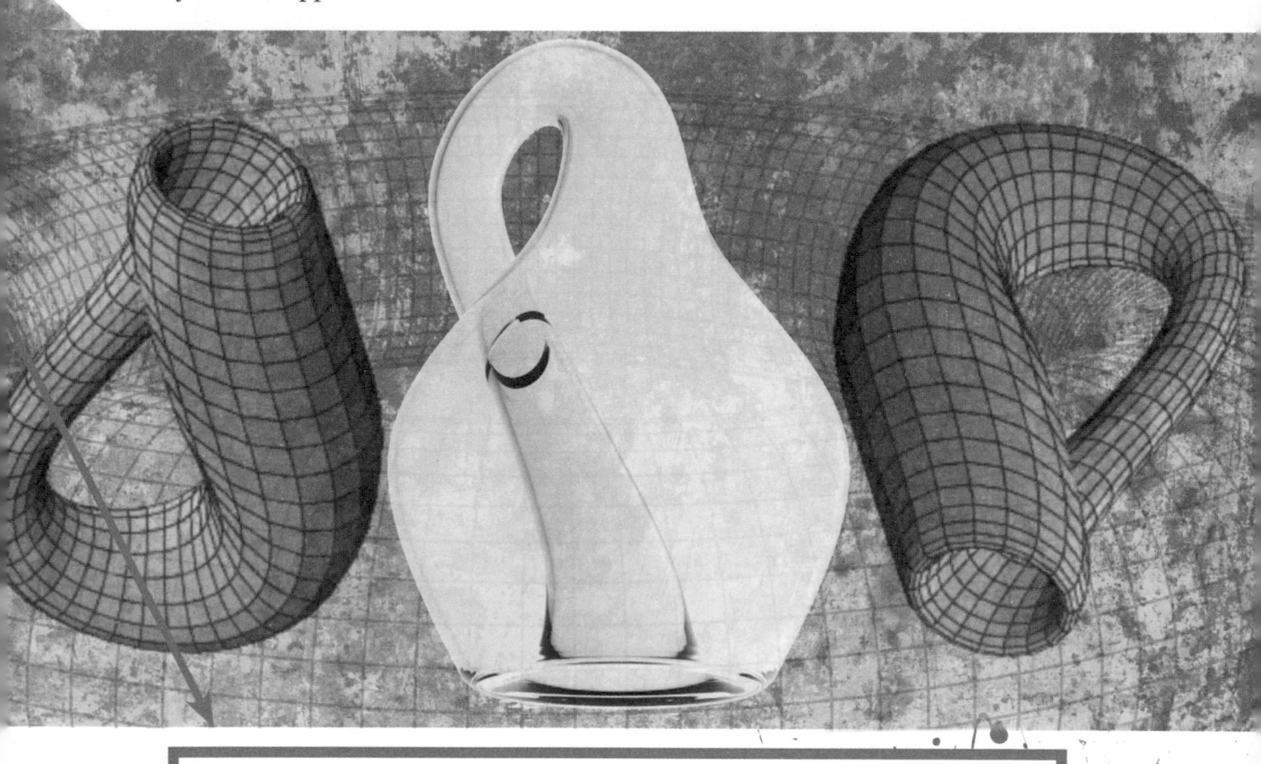

3/Hack:

Topology studies space from a different perspective from geometry. Instead of measuring parts of it, it tries to show us how the whole space is connected together.

Topologists do not care about the difference between triangles and squares or between doughnuts and coffee cups.

No.71 Triangulation

Complexes of simple things

1/ Helicopter view: The simplest unit of zero-**dimensional** space is a point. The equivalent for one-dimensional space is a line segment (a finite, straight length with a point at each end). Think of this as taking two points and filling in the one-dimensional space between them.

In two **dimensions** the simplest unit is a triangle, which is a region of space bounded by three lines joined at their end-points. In three dimensions, the equivalent is a tetrahedron, which is made by joining four triangles together along their edges and filling in the interior. You can probably work out the pattern for higher dimensions, though we can no longer make pictures of them.

Each of these is called a *simplex*. If we join several of these together we get a *simplicial complex*, which can be thought of as dividing up the space it occupies into the simplest possible parts: *triangulating* it.

Dividing 2D space into triangles or 3D space into tetrahedra helps us find our way around.

2/ Shortcut: We usually allow *simplices* to be distorted as if made of rubber—so it is OK for a triangle to be stretched out into a circle or pinched into a square. Similarly, straight lines can become curved and a tetrahedron can be inflated into a ball.

The point of simplices is that they are simple, and simplicial complexes can be broken down into them. So *triangulating* a space is a great way to get to grips with it. It is a basic tool in modern geometry and **topology**.

See also//
70 Topology, p.144
72 The Euler Characteristic, p.148
76 Dimensions, p.156
80 The Regular Tilings, p.164
84 Algebraic Topology, p.172
94 Graphs, p.192

3/ Hack: A simplex is a basic unit of space. A simplicial complex is several simplices glued together.

Triangulating a space gives us a new, more topological perspective on it.

No.72 The Euler Characteristic

Leonhard Euler (1707–83)

Probing topology with triangulations

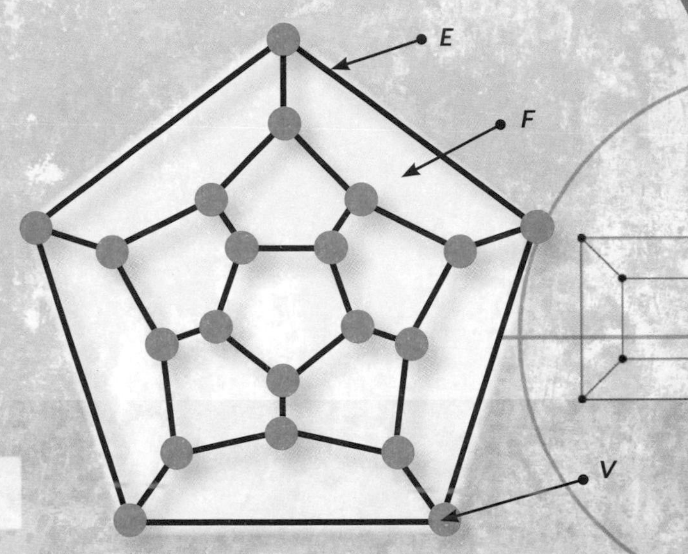

1/ Helicopter view: There are endless ways to **triangulate** a space. To convince yourself, take a piece of paper and draw intersecting lines until it is divided into triangles. Now ask yourself how many different patterns could be made that way. Certainly the answer is infinite.

It would be surprising if we could find anything that all these triangulations have in common. The *Euler characteristic* is such a thing: a number that is easy to calculate from any one triangulation and is the same no matter which you use.

Another surprise, though, is that the number changes depending on the underlying space the triangulation lives in. This has nothing to do with whether the space is big or small, curved or straight, and so on. Rather, it depends on how the space as a whole is connected together. In other words, it depends only on its **topology**.

Graphs have the same *V*, *F*, and *E* on a plane or a sphere because they have the same Euler characteristic.

2/Shortcut: Triangulate a surface and count how many triangular faces there are and call that F. Now count how many triangle edges there are and call that E. Finally, count the vertices (that is, triangle corners) and call that V. Then the *Euler characteristic* is just $V - E + F$.

Triangulations of higher-**dimensional** spaces have higher-dimensional **simplices** but there's a simple generalization of the formula that works for those as well.

See also//

70 Topology, p.144

71 Triangulation, p.146

76 Dimensions, p.156

84 Algebraic Topology, p.172

94 Graphs, p.192

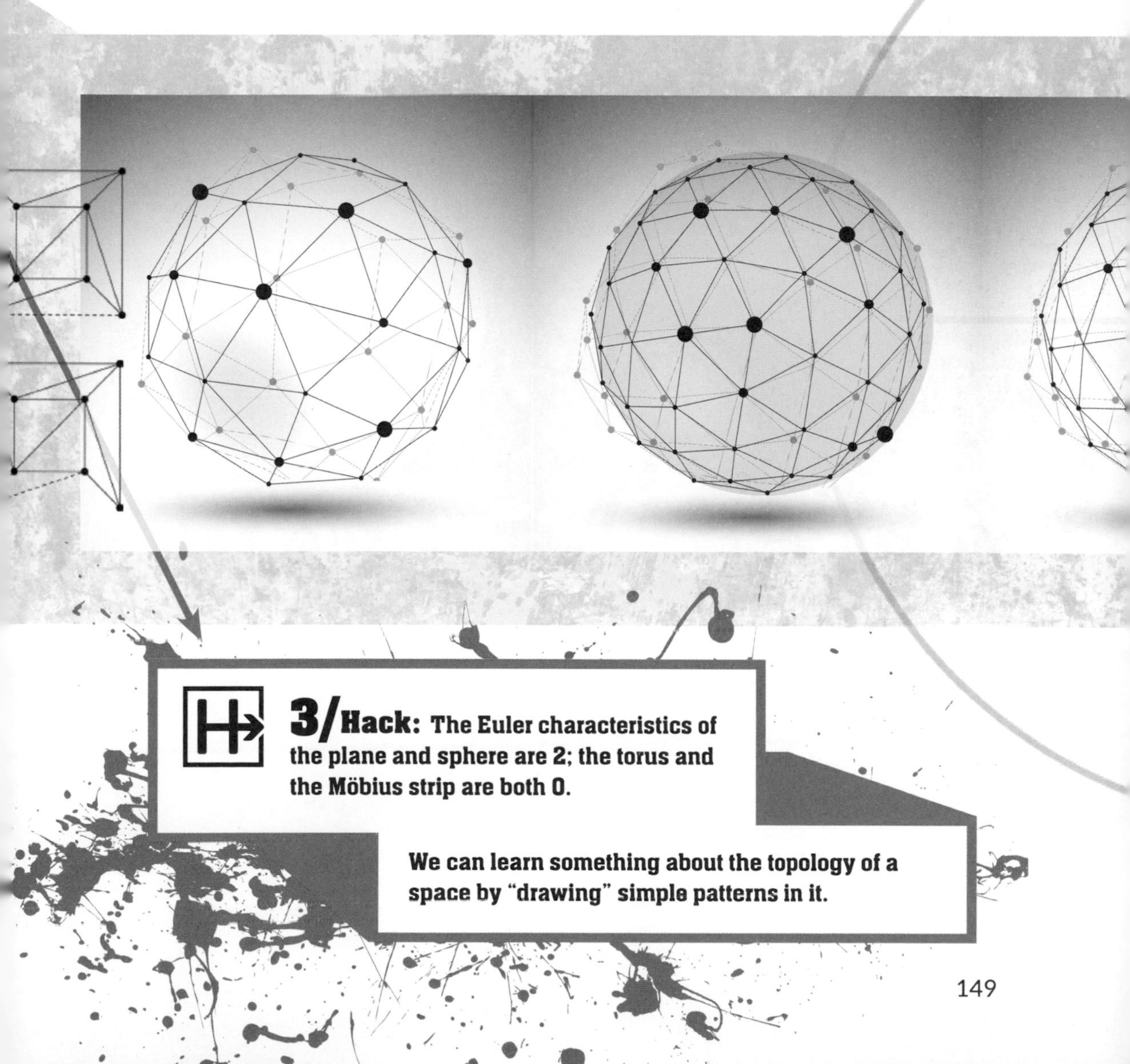

3/Hack: The Euler characteristics of the plane and sphere are 2; the torus and the Möbius strip are both 0.

We can learn something about the topology of a space by "drawing" simple patterns in it.

No.73 The Illumination Problem

A hall of mirrors challenges the geometers

Roger Penrose (b.1931) has made many contributions to mathematics, including publishing the first solution to the illumination problem in 1958.

1/ Helicopter view: Suppose you are in a room whose walls are completely covered in mirrored glass, without windows or doors. You have a single light bulb. Are any parts of the room in shadow?

Light travels in straight lines and bounces off a mirror just like a billiard ball. Unlike the ball, though, a light particle never slows down, so every single ray the bulb gives out just keeps bouncing around the room forever. Since light is being emitted in a constant stream in all directions, it seems reasonable to assume that the room will be completely flooded with light.

The shape of the room turns out to be crucial. In 1958 Roger Penrose designed a room with curved walls that always has a dark region no matter where you put the bulb. In 1995 George Tokarsky found an example with straight walls, although the unlit regions can be reduced to a collection of individual points, which you might feel is not the same thing.

Right: A single point of light placed anywhere inside George Tokarsky's room leaves only a single point in darkness.

2/ Shortcut: The *Illumination Problem* is remarkable for several reasons. First, like the **Collatz Conjecture**, it is easy to understand without any complicated mathematics, but it appears that nobody thought of it until quite recently. Second, the solutions we have so far do not use very fancy methods either.

If nothing else, it demonstrates two things. First, even in quite elementary geometry there are probably basic questions nobody has thought of asking yet. Second, just because a question is new does not mean it needs advanced mathematics to answer it.

See also//

15 The Collatz Conjecture, p.34

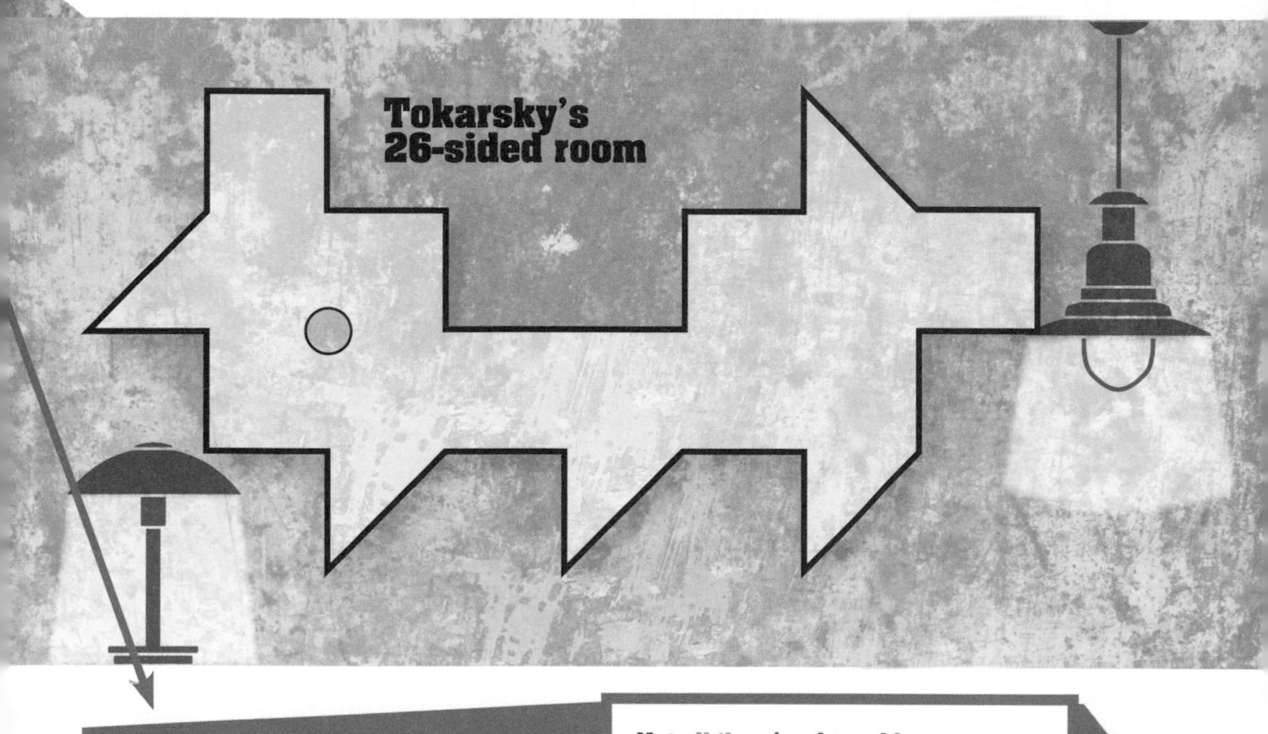

3/ Hack: You can design a room with completely mirrored walls that will have shadowy regions no matter where you put a single source of light.

Not all the simple problems have been discovered yet, and the solutions of new problems need not involve new mathematical ideas.

No.74 Metric Spaces

Do not take measurement for granted

1/ Helicopter view: Early on in our education we learn about length. We measure it in yards, feet, and inches, meters and centimeters, or other such units. We learn to use a ruler to measure the lengths of straight lines, and perhaps a tape or piece of string to do the same for curved ones. Later we encounter the concepts of area and volume, and we learn to calculate them using lengths of lines as the starting point.

We can also calculate the distance between two points, assuming it is the length of the straight line between them. But this assumption might not always hold. For example, in a city you often cannot travel in a straight line from point to point, but must follow the lines of the streets instead. Here distances must be calculated differently.

The idea that there might be multiple ways to calculate distance leads to the abstract notion of a *metric*.

The distance from A to B depends on how you get there. Sometimes a straight line gives the wrong answer.

2/Shortcut:

Suppose we represent a space abstractly by a set of “points” (whatever they might be). It becomes a *metric space* when we add a **map** that measures the distance between points. It does this by taking any pair of points and mapping them to a **real number**.

A metric is subject to a few common-sense constraints.

$d(p, p) = 0$: the distance from a point to itself is zero.

$d(p, q) = d(q, p)$: the distance from p to q is identical to the distance from q to p.

$d(p, q) \leq d(p, r) + d(r, q)$: this—the triangle inequality—says that going from p to q via some other point r is never quicker than going directly.

See also//
9 Maps, p.22
26 Real Numbers, p.56
59 Euclidean Spaces, p.122
78 Spherical Geometry, p.160
79 Hyperbolic Geometry, p.162

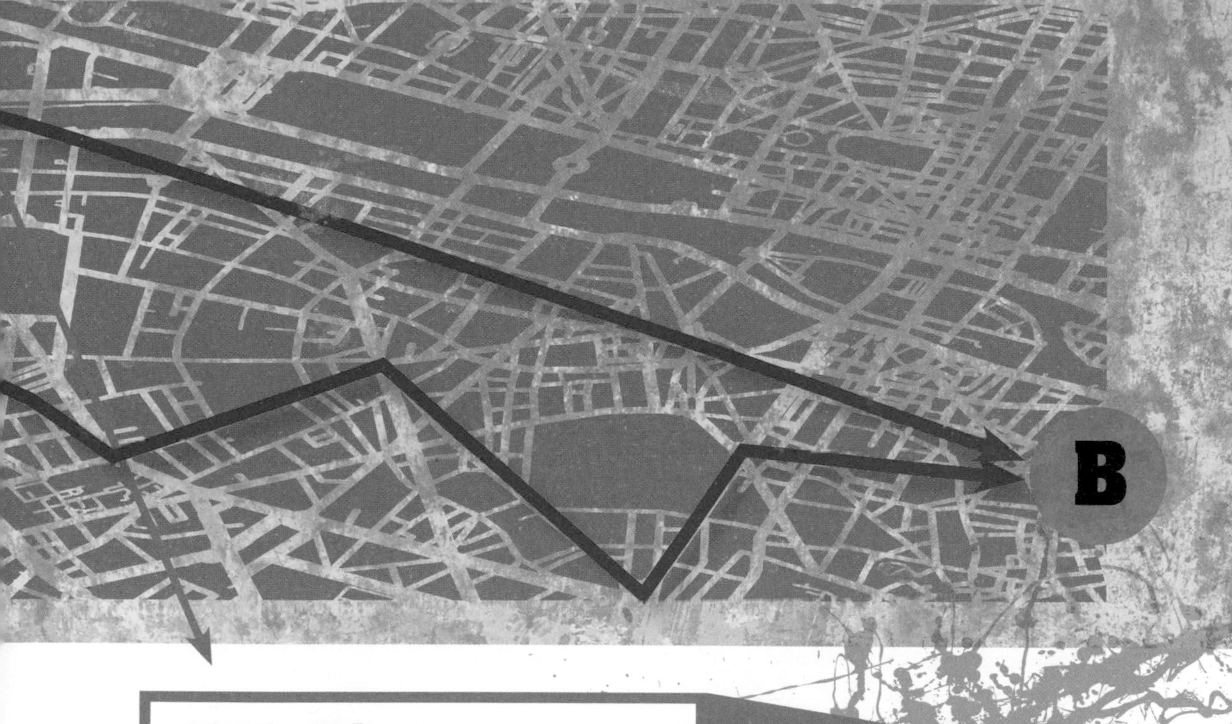

3/Hack:

A metric is a map that tells us how to calculate distances between any pair of points in a space.

A metric can be anything at all that meets the three criteria.

No.75 Curvature How space changes

1/ Helicopter view: We all have an intuitive idea of how curved a line is at a given point, for example, how hard you would have to turn the steering wheel to drive a car around a bend at that point. Imagine you were driving along the line and it suddenly vanished. If you did not move the steering wheel, you would be left turning in a circle.

We call this the *osculating circle* at that point. If the circle is big, the bend was not very tight. Conversely, a small circle means a lot of curvature. So we use the number $1/r$ to represent the curvature of the line at that point, where r is the radius of the osculating circle. It can be positive or negative, depending on the direction of the curve. Since it involves measuring lengths, it requires a **metric**.

Extending this to surfaces and spaces of higher **dimensions** found a dramatic application in Albert Einstein's theory of relativity, which models gravity as *curvature* of spacetime.

The osculating circle at a point defines the curvature of a line. This extends to higher-dimensional objects, such as surfaces.

2/ Shortcut: The *curvature* of a two-**dimensional** surface at a point is calculated by standing at that point and looking around. Look for the direction where walking in a straight line gives the biggest curvature; call this K_{max}. Do the same for the smallest curvature; call this K_{min}. Multiplying these gives the Gaussian curvature there. If K_{max} and K_{min} have different signs, the curvature will be negative.

Curvature can change from point to point, and a **tensor field** is the right sort of object to represent it.

See also//
60 Manifolds, p.124
65 Tensor Fields, p.134
74 Metric Spaces, p.152
76 Dimensions, p.156
79 Hyperbolic Geometry, p.162

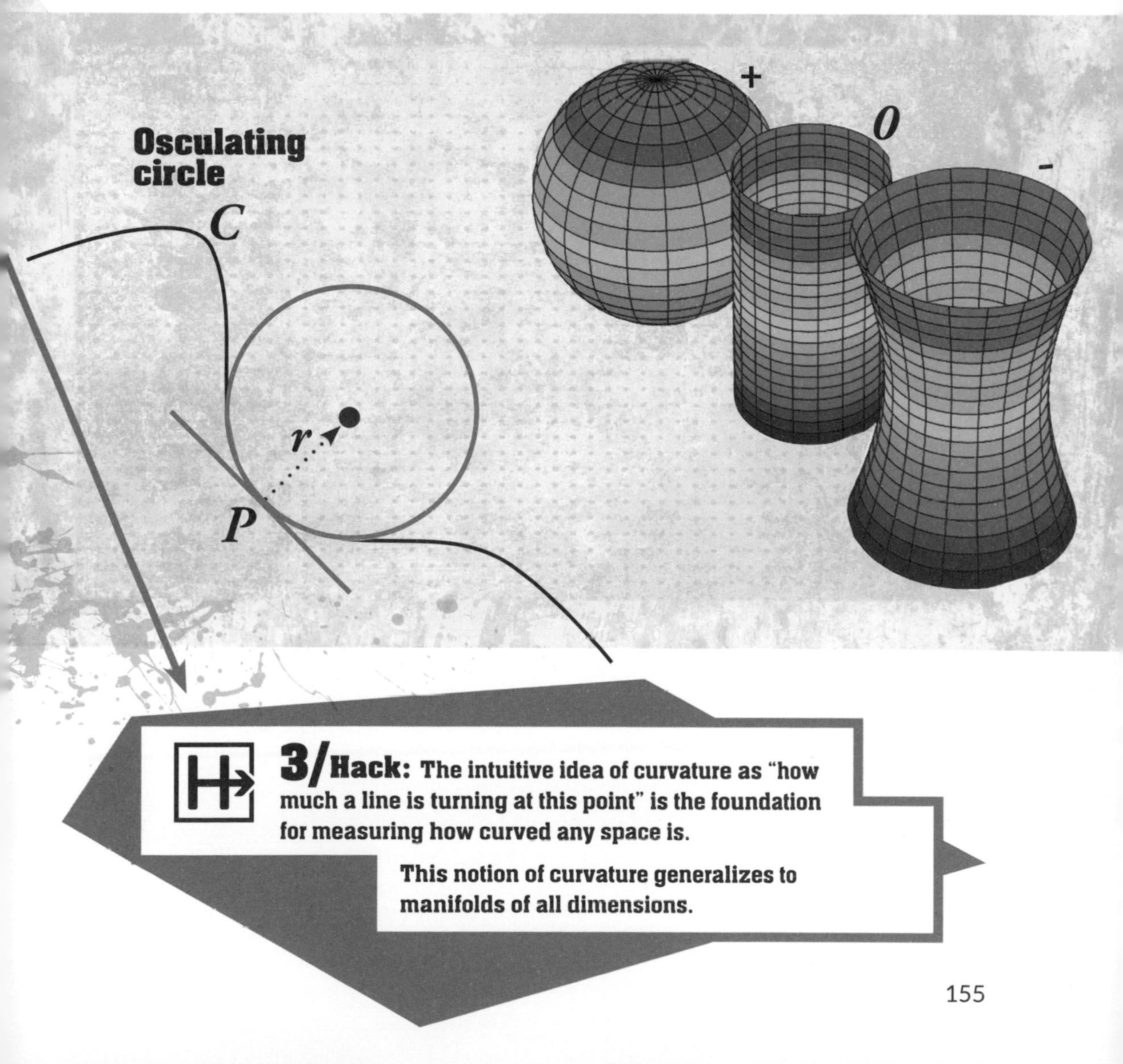

3/ Hack: The intuitive idea of curvature as "how much a line is turning at this point" is the foundation for measuring how curved any space is.

This notion of curvature generalizes to manifolds of all dimensions.

No.76 Dimensions Coordinating space

1/ Helicopter view: Here is a way to get oriented in a two-*dimensional* (2D) space. Mark one point as special (it does not matter which). Call it "the origin." Every other point in the space can be identified by drawing a straight line from the origin to that point. Put an arrow at the end touching the target point and you can call it a **vector**.

The next step is to choose two vectors that we can use to describe all the others. One (call it E) might point 1km east of the origin. The other (call it N) might point 1km north instead. We can now describe any point as being "x km east and y km north" (note that x and y can be negative!).

That is, for every vector v, $v = xE + yN$. The properties of vector spaces ensure this works. We say E and N together form a *basis* for the space.

Below: A 2D space needs two coordinates to determine a point. *Right:* We need six numbers to describe the movement of an aircraft, making it a point in 6D space.

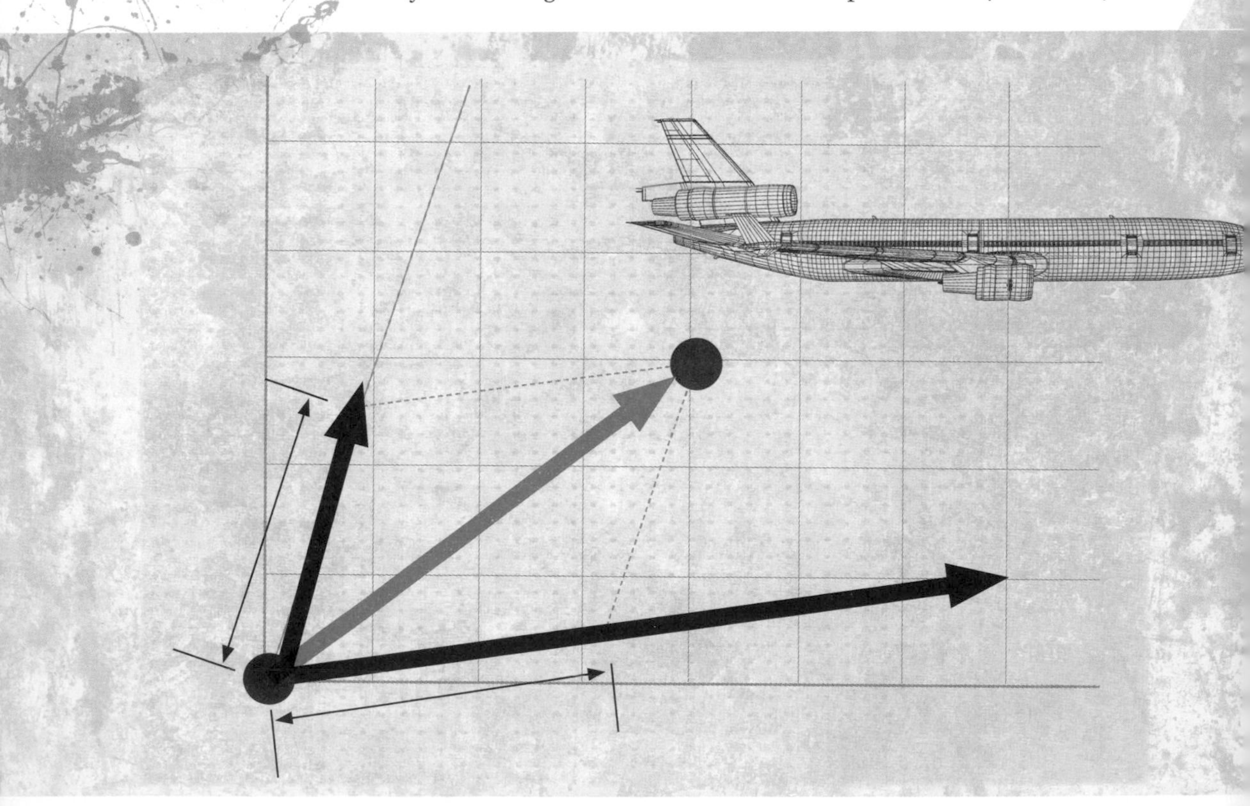

2/Shortcut: If the basis contains exactly two vectors, the space is 2D *by definition*. We could choose different vectors but we would always need two for this particular space.

In 3D I might use "up," "left," and "forward," with myself as the origin. In a different context, "latitude," "longitude," and "altitude" form a basis, with Greenwich (London) at sea level as the origin.

A flying plane has three independent directions of movement and three of rotation; thus its possible motions can be thought of as points in a 6D space.

See also//
57 Vectors, p.118

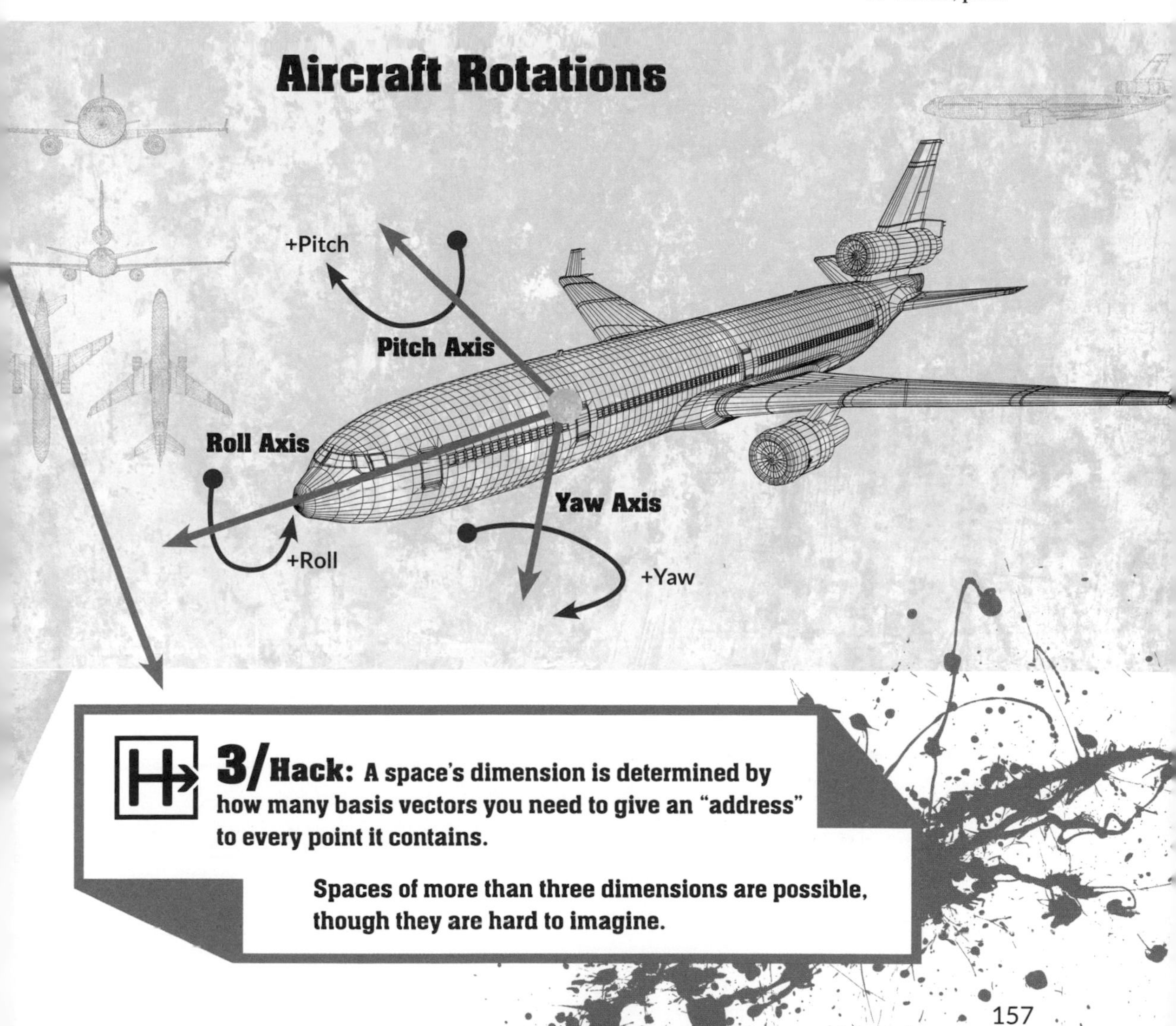

3/Hack: A space's dimension is determined by how many basis vectors you need to give an "address" to every point it contains.

Spaces of more than three dimensions are possible, though they are hard to imagine.

No.77 Fractional Dimensions

What is in between a length and an area?

Benoît Mandelbrot (1924-2010) gave fractals their name and did much to popularize them.

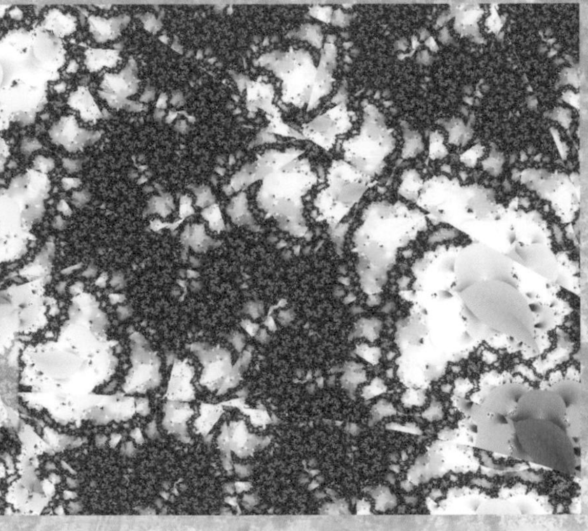

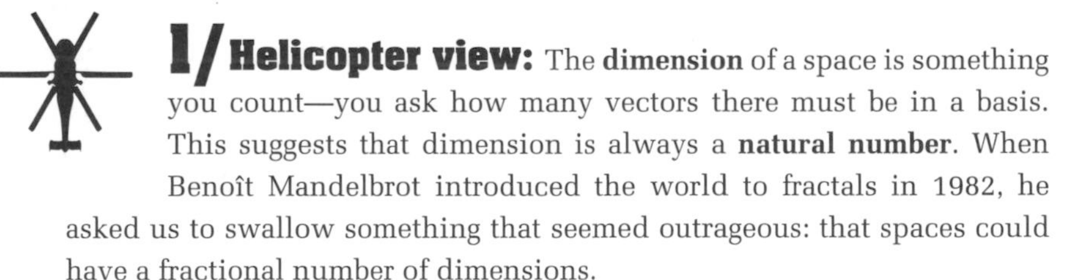

1/ Helicopter view: The **dimension** of a space is something you count—you ask how many vectors there must be in a basis. This suggests that dimension is always a **natural number**. When Benoît Mandelbrot introduced the world to fractals in 1982, he asked us to swallow something that seemed outrageous: that spaces could have a fractional number of dimensions.

The term "fractal" is imprecise, but fractals are complex never-ending patterns in which the same pattern appears at smaller and smaller scales. They often have strange **metrical** properties. For example, the Koch Snowflake is an infinitely long fence that encloses a small garden. The Cantor set is a kind of dust of line segments too short to have length but more than mere points.

The Cantor set seems like more than a zero-dimensional set of points, but less than a one-dimensional line, while the Koch Snowflake is more than a line but less than an area, and so on. A fractal need not have a *fractional dimension*. In fact, the famous Mandelbrot set does not.

Right: The Menger sponge and the Koch snowflake.

2/Shortcut: Of course, this requires a new understanding of what "**dimension**" means. There are various methods for calculating *fractal dimension*, and all rely on taking **limits** of infinite processes. One of the most widely used is called the Hausdorff dimension.

The Hausdorff dimension of the Cantor set is about 0.63. The Koch Snowflake's is about 1.26. The Menger Sponge, which seems to sit in between the notions of area and volume, has a Hausdorff dimension of about 2.73. All these correspond to our sense that they may be "in between" the more familiar dimensions.

See also//
4 Limits, p.12
14 Natural Numbers, p.32
74 Metric Spaces, p.152
76 Dimensions, p.156

3/Hack: The study of fractals led to new ideas about dimensions in geometry that allow in-between dimensions to exist and be calculated precisely.

Such objects blur the boundaries between dimensions.

No.78 Spherical Geometry

As above, so below

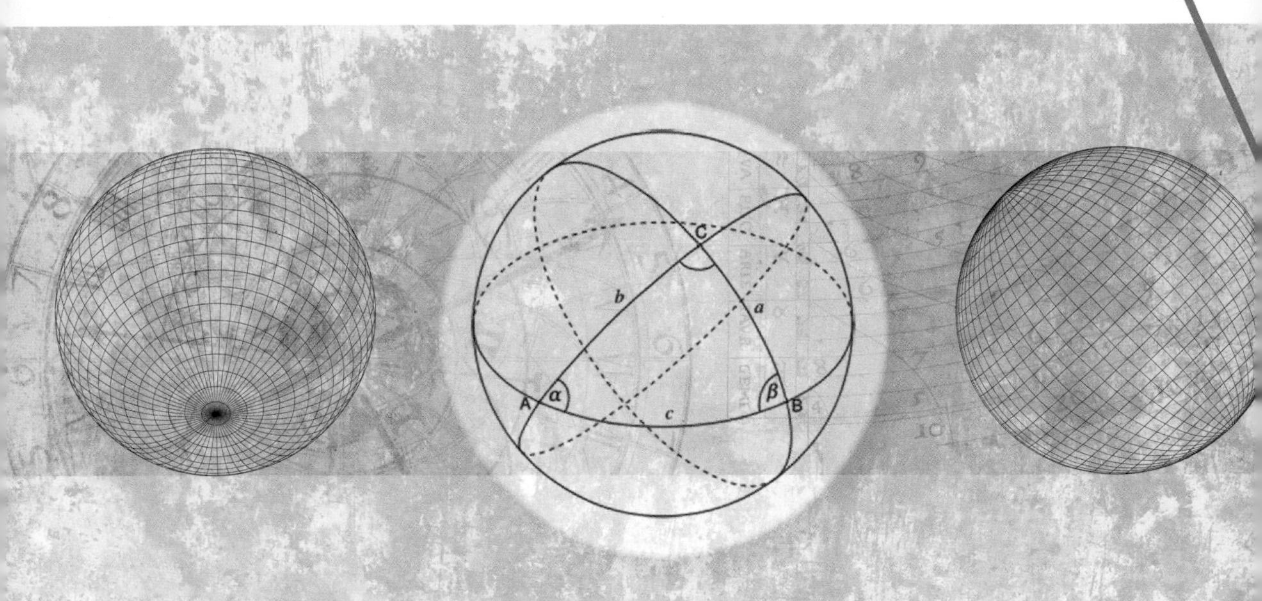

Straight lines on a sphere look curved from the outside; spherical triangles have fatter angles than planar ones.

1/Helicopter view: When ancient peoples looked up at the sky they could be forgiven for thinking the stars and planets are embedded in a huge dome, like the inside of half a sphere. Astronomical observations of this "dome" led to the development of geometric rules for how moving points, lines, triangles, and so on behave on spheres.

What's more, the Earth had been known to be roughly spherical in antiquity. This became a practical concern when, in the 16th century, it became necessary to chart routes across the Atlantic. In the absence of landmarks sailors had to rely heavily on flat maps, which involved understanding precisely how they distort the *spherical geometry* of the Earth.

However, it was only at the end of the 18th century that mathematicians realized the significance of these practical considerations. Spherical geometry came to be seen as its own domain, distinct from the flat geometry of **Euclidean space**.

2/Shortcut: *Spherical geometry* has the same **axioms** as **Euclidean** geometry except for one; the Fifth Postulate concerning **parallel lines** is rejected.

On the surface of a sphere, a "straight line" is a circle whose center is the center of the sphere (on the Earth, think of the equator or any line of longitude). Two "straight lines" of this kind always intersect at two points on the surface of the sphere, meaning there are no parallel lines!

Changing this single axiom has far-reaching consequences, making the resulting geometry quite different.

See also//
1 Axiom, Theorem, Proof, p.6
59 Euclidean Spaces, p.122
68 Parallel Lines, p.140
74 Metric Spaces, p.152
79 Hyperbolic Geometry, p.162

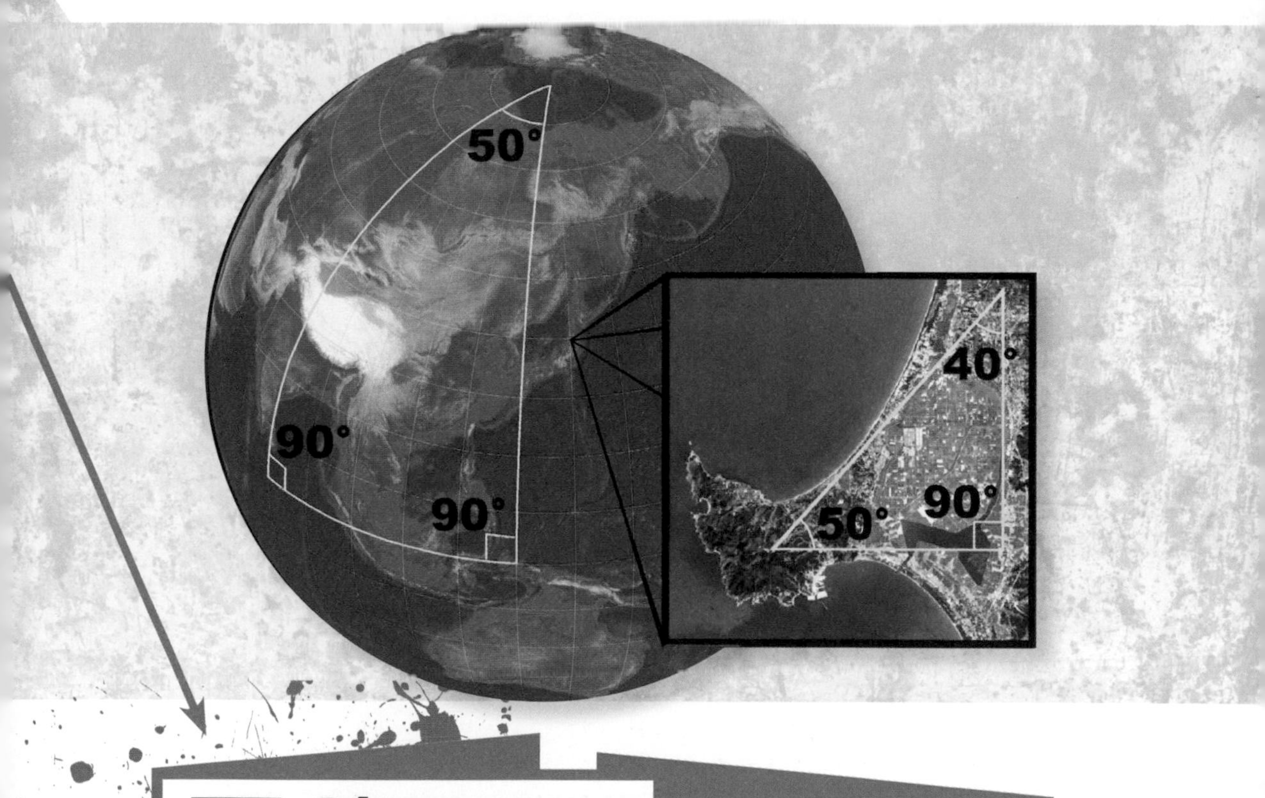

3/Hack: **On a sphere the shortest path between two points is part of a circle. A more technical way to say this is that we have introduced a new metric.**

Spherical geometry is Euclidean geometry without any parallel lines.

No.79 Hyperbolic Geometry

When curvature goes negative

1/ Helicopter view: Euclid's Fifth Postulate (**axiom**) amounts to this: given a straight line and a point not on the line, there is exactly one parallel straight line that goes through the point.

Spherical geometry comes from rejecting this postulate in one way: there is *no* such parallel line.

Hyperbolic geometry comes from rejecting the postulate in the opposite way: there are an *infinite number* of parallel lines. This was a scandalous idea when first proposed, and it is said that János Bolyai was discouraged from publishing these ideas to preserve his reputation.

Consider a **manifold** of any **dimension**. **Euclidean** geometry is true at points where the **curvature** is zero. Spherical geometry holds where the curvature is positive. Hyperbolic geometry is the right one for points of negative curvature.

Although curvature can change as we move around in a space, it is easiest to think about spaces where it stays the same at every point.

Regions of negative curvature, which slope both up and down depending on which way you look, are a common feature of contemporary architecture.

2/Shortcut: 2D **Euclidean** and **spherical geometries** can be realized as the flat plane and the sphere. There is no easy-to-picture surface with constant negative **curvature**, however, so *hyperbolic geometry* has to rely on imperfect visual models.

For surfaces—2D **manifolds**, that is—these three geometries are the only ones possible. At every point on a surface, after all, the curvature must be positive, negative, or zero, making the geometry spherical, Euclidean, or hyperbolic, respectively. The situation in three dimensions is considerably more complicated.

See also//

1 Axiom, Theorem, Proof, p.6
59 Euclidean Spaces, p.122
60 Manifolds, p.124
75 Curvature, p.154
76 Dimensions, p.156
78 Spherical Geometry, p.160

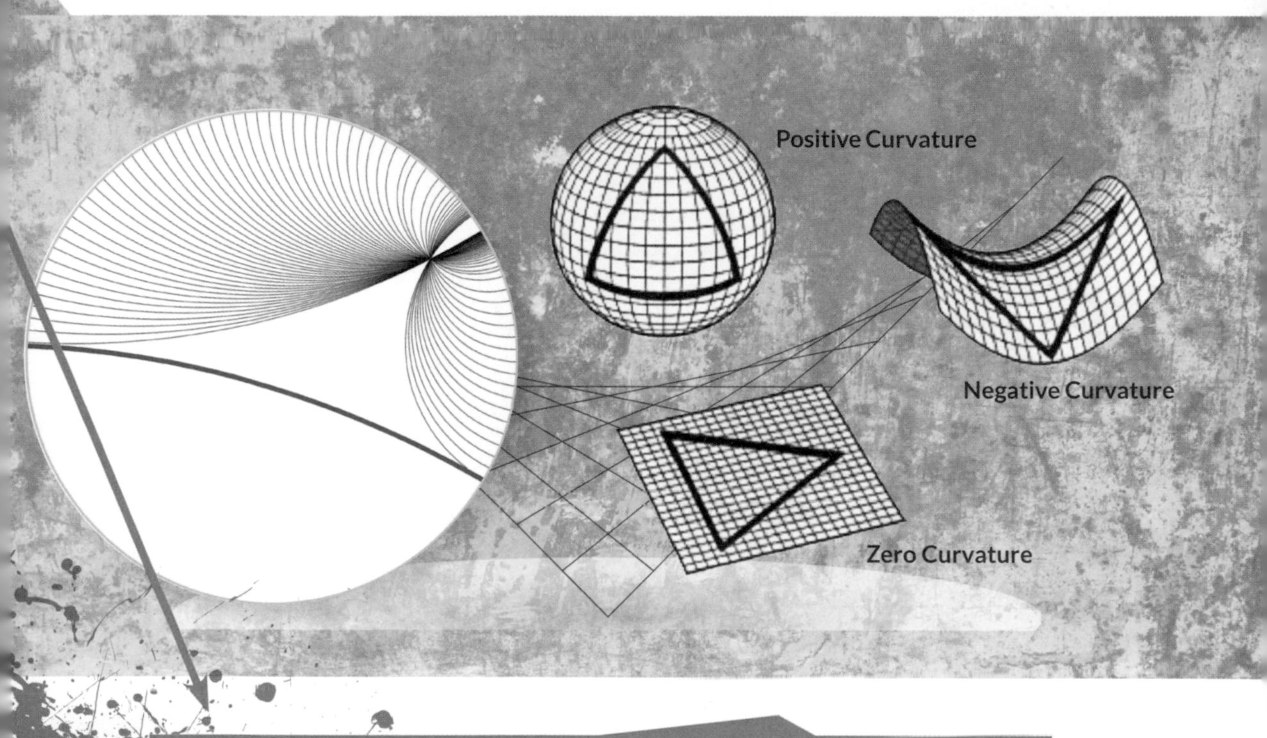

3/Hack: Hyperbolic geometry is the same as Euclidean, except that parallel lines are not unique. Given a line, there are an infinite number of lines that are parallel to it through any point.

Hyperbolic geometry is true in regions of space with negative curvature.

No.80 The Regular Tilings

Pixelating space

1/ Helicopter view: Imagine covering a wall with tiles. How many ways could you do it? Well, there are lots of ways but suppose you are limited by some constraints. First, all your tiles are identical regular polygons, meaning all their sides and angles are the same. Second, the tiles must be laid edge-to-edge, never corner-to-edge like bricks in a wall.

Perhaps surprisingly, there are only three possibilities. You can tile your wall with triangles, squares, or hexagons, and there is only one way to do each one. No others will work.

At least that is true if you are tiling a flat plane. What if we tried a sphere instead? Then there are five ways to do it. Three involve triangles, one uses squares, and the last uses pentagons. This is because s**pherical geometry** is not the same as Euclidean.

Finally, if your surface is **hyperbolic**—that is, if it has negative **curvature**—there are an infinite number of regular tilings.

The three regular tilings of the plane were known to Euclid and often appear in decorative patterns, such as these tiles from the Alhambra palace.

2/Shortcut: These simple facts are a consequence of the **curvature** of the space. Curvature affects geometric measurements like lengths, angles, and areas, and this isn't very surprising. The fact that new *kinds* of tiling exist in these geometries is more dramatic, though, and suggests these spaces really are radically different from one another.

Tilings have higher-dimensional analogs. Consider the way a crystal is built by stacking together identical geometric units. These divide up the space in a regular way, too.

See also//
71 Triangulation, p.146
75 Curvature, p.154
78 Spherical Geometry, p.160
79 Hyperbolic Geometry, p.162
81 Thurston's Geometrization Theorem, p.166

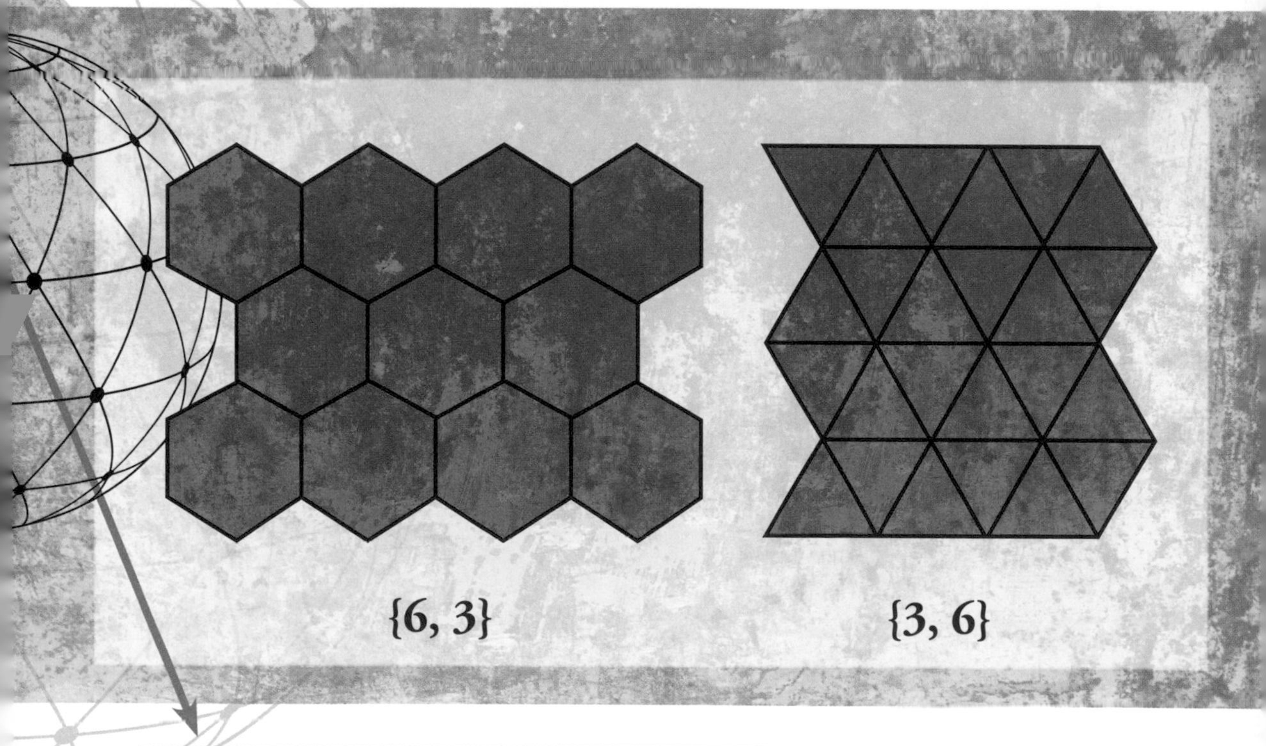

3/Hack: There are three regular tilings of the plane, five of the sphere, and an infinite number in hyperbolic space.

A regular tiling uses one tile that is a regular polygon, with tiles laid edge-to-edge and corner-to-corner.

No.81 Thurston's Geometrization Theorem

What 3D spaces look like locally

1/ Helicopter view: The three kinds of **curvature** a surface can have give rise to three radically different geometries. But a surface is merely a two-dimensional thing. What about in higher dimensions?

The extension to three dimensions was proposed by William Thurston in 1982 and sparked a two-decade research program. It was finally resolved (in the way Thurston had predicted) in 2003 by Grigori Perelman as part of his proof of the **Poincaré Conjecture**. Perelman was offered a million-dollar prize for his work by the Clay Foundation, which he famously declined.

This result tells us that there are eight fundamentally different geometries a region of a three-dimensional **manifold** can have. Three of the eight are higher-dimensional versions of the geometries we find in two dimensions; the rest are difficult to form any intuitive picture about.

A 3D space hyperbolic can be tiled with regular dodecahedra.

2/Shortcut: The geometries are, in approximate order of weirdness:

E^3, which is flat ordinary 3D space

S^3, which is **spherical**

H^3, which is **hyperbolic**

$S^2 \times \mathbb{R}$, a product of the sphere with the real number line

$H^2 \times \mathbb{R}$, a product of 2D hyperbolic space with the real number line

$SL_2(\mathbb{R})$, which comes from a certain class of 2 x 2 **matrices**

Nil geometry

Sol geometry.

The definitions are highly technical. Even drawing pictures is almost impossible, so abstract methods are best for discovering things about these geometries.

See also//
60 Manifolds, p.124
63 Matrices, p.130
75 Curvature, p.154
78 Spherical Geometry, p.160
79 Hyperbolic geometry, p.162
80 The Regular Tilings, p.164
86 The Poincaré Conjecture, p.176

3/Hack: Any region of a 3D manifold must be governed by one of eight geometries.

Geometrically speaking, 3D space is much richer than 2D space, which only has three geometries.

No.82 Projective Geometry

The mathematics that painters discovered

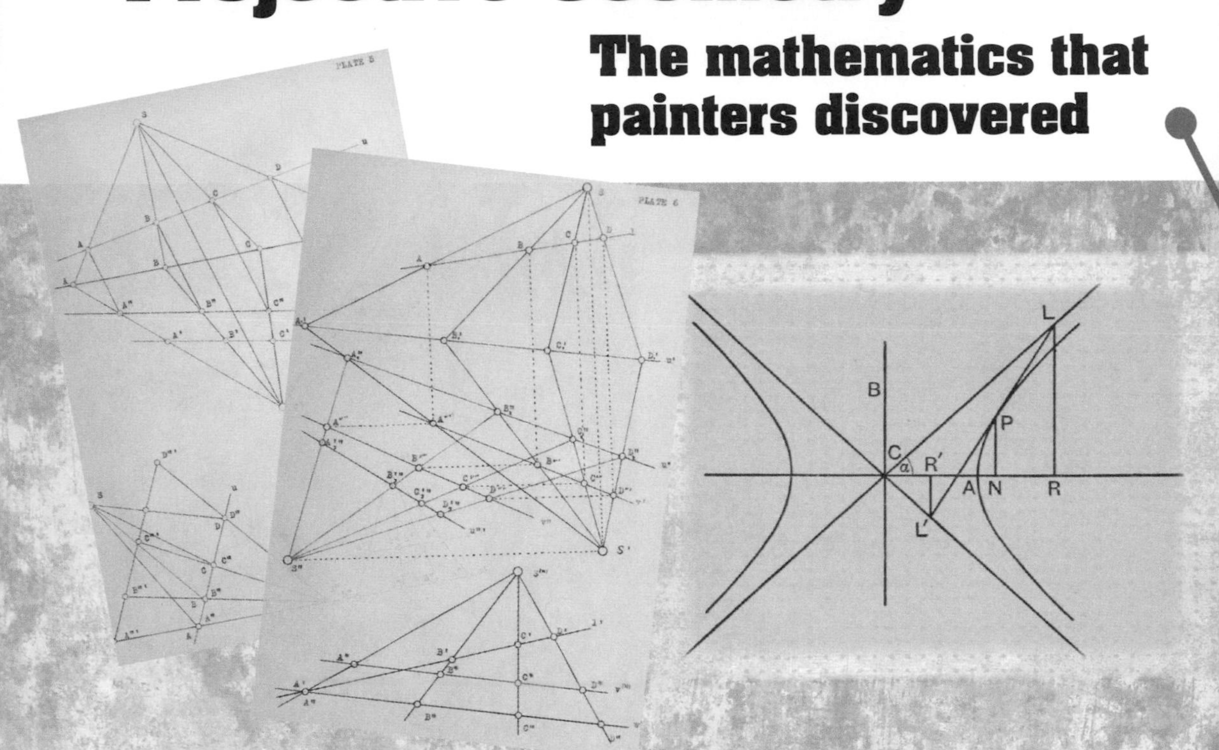

1/ Helicopter view: The art of rigorous perspective drawing developed in Florence, Italy, in the early 1400s. Previous artists had some *ad hoc* ways to give an impression of depth, but perspective drawing is more than that. It is a system of lines that tell you, in theory, exactly where everything in your picture should go.

The picture is a 2D representation of a 3D world, and perspective gives you a way to consistently "project" the world onto the picture along straight lines. These can be thought of as representing rays of light coming from objects in the 3D space and hitting your eye, passing through the artist's picture on the way.

Perspective was rapidly perfected as a practical tool for making visual representations. It was not until the 1600s that mathematicians began to take a serious interest in it, and it took a long time to be fully appreciated. Today it is considered a cornerstone of all of geometry.

Above: The techniques of perspective drawing contain a distinctive geometry.
Right: Drawing by Albrecht Dürer (1471–1528) of a machine that carries out perspective projection.

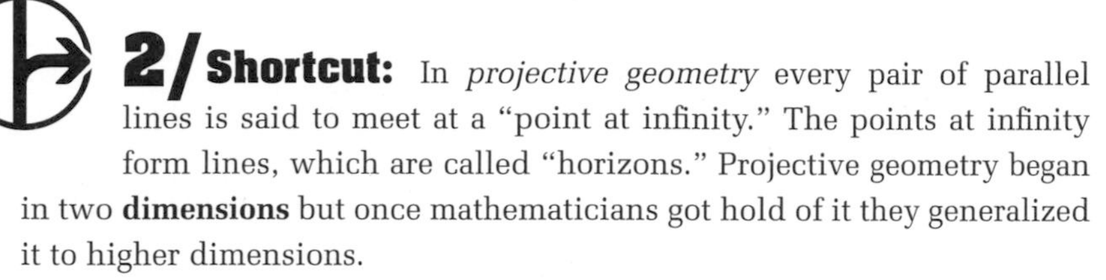

2/ Shortcut: In *projective geometry* every pair of parallel lines is said to meet at a "point at infinity." The points at infinity form lines, which are called "horizons." Projective geometry began in two **dimensions** but once mathematicians got hold of it they generalized it to higher dimensions.

The main **axioms** of projective geometry are shockingly simple:

- Every two distinct points lie on a unique line.
- Every two distinct lines meet at a unique point.

From these (and another **axiom** to rule out silly examples), everything else follows.

See also//

1 Axiom, Theorem, Proof, p.6

76 Dimensions, p.156

78 Spherical Geometry, p.160

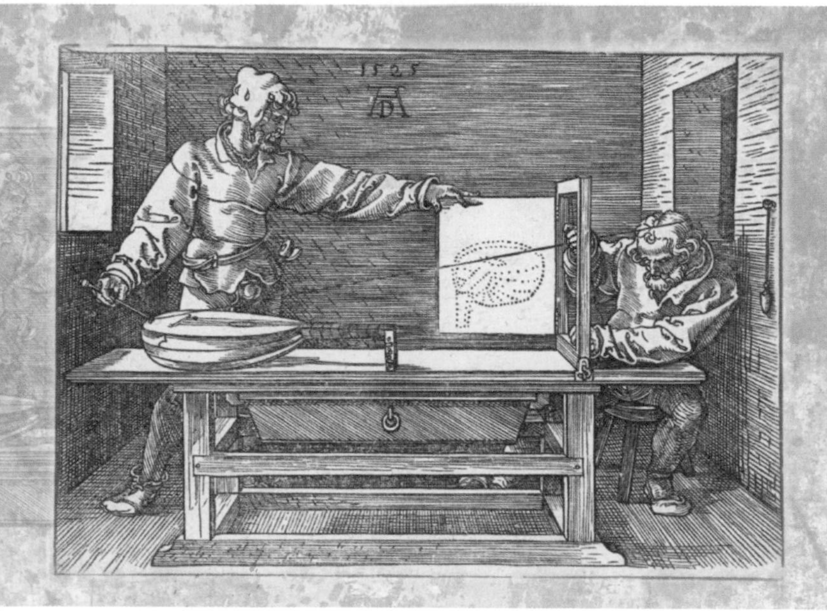

3/ Hack: The rules for rigorous perspective drawing involve projecting 3D space into 2D. This gives rise to a special kind of geometry.

Projective geometry generalizes the idea that any *point* you see is the result of a *ray* of light hitting your eye.

No.83 The Tesseract

A four-dimensional cube

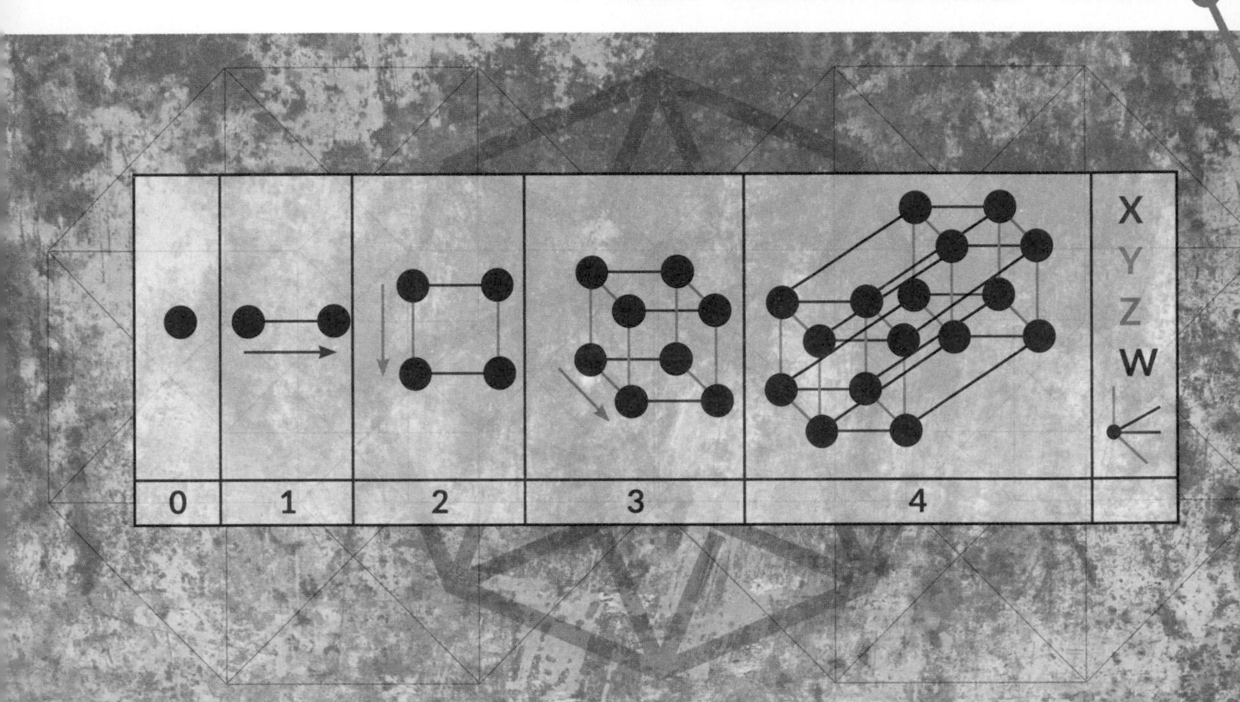

1/ Helicopter view: Mark a point somewhere and push it along, say 1 meter eastward. This traces a line segment. Now push this segment 1 meter northward. You have made a square, lying flat on the ground. Now pull the square 1 meter upward and you have a cube. Notice that each push or pull was at right angles to the previous ones.

That is as far as we can take things in 3D space because we have used up all three **dimensions**. If we had another dimension, though, we could repeat it again, pushing the cube in a direction that is at right angles to it, through our new fourth dimension. This traces out a *tesseract*.

Alternatively, remember you can make the boundary of a (3D) cube by gluing together six (2D) squares along their (1D) edges. Similarly, there is a way to make a (4D) tesseract by gluing together eight (3D) cubes on their (2D) faces.

Several ways to visualize a tesseract, including Le Grande Arche de la Défense in Paris.

2/ Shortcut: It's hard to imagine most higher-**dimensional** shapes. The *tesseract* was one of the first to be described in ways we have a chance to grasp intuitively. We can divide up a 2D surface into squares, a 3D space into cubes, and a 4D space into tesseracts, so this is really helpful.

The tesseract construction extends to higher **dimensions** in the obvious way; push a tesseract along a fifth dimension to create the 5D analog of a cube, and so on.

See also//
71 Triangulation, p.146
76 Dimensions, p.156
86 The Poincaré Conjecture, p.176

3/ Hack: Higher-dimensional spaces are tough to imagine. Simple ways to break them down like this really help.

Going up in dimensions we have a line, a square, a cube, and a tesseract. We can continue as far as we need.

No.84 Algebraic Topology

Magic lanterns for mathematicians

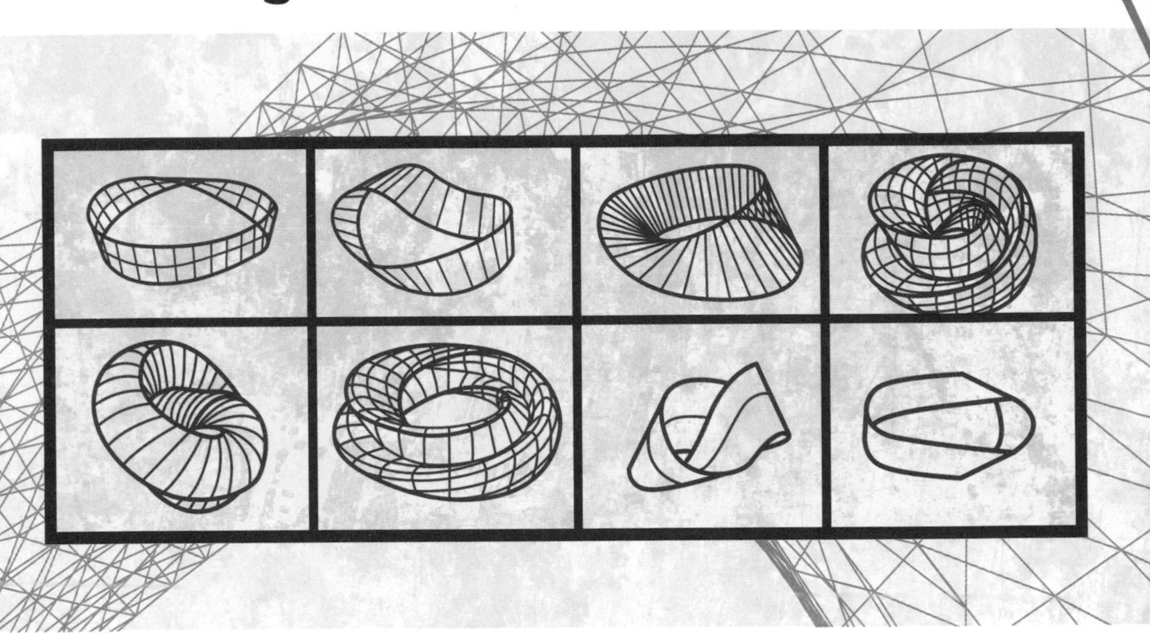

1/ Helicopter view: **Topology** is hard. It studies strange spaces that are unlike anything we can possibly experience. They might have high **dimensions** and be connected together weirdly, for example. You can't just look at a seven-dimensional **manifold** and make deductions about it. Topologists need tools that help them.

One of the most powerful involves attaching a simpler object to the space you are trying to study and letting its structure tell you something about the space. Often that object is a **group**.

For example, imagine picking a point and creating loops (like pieces of string) that start and end there. You consider two loops to be basically the same if one can be stretched and pulled into the same shape as the other without leaving the space.

With a little trickery, these loops can be turned into an algebraic object, the space's "fundamental group," which is good at finding holes you did not know were there.

Above: Various surfaces, each topologically either a Klein bottle or Möbius strip. Both are single-sided surfaces but only the Möbius strip has an edge.
Right: Algebraic topology creates simpler images of complicated objects.

2/Shortcut: Topologist Allen Hatcher likened *algebraic topology* to a magic lantern that shines a light through a topological space to produce a projected image, which is the algebraic structure. We can then deduce things about the space from the image it projects.

Two other groups used to study topological spaces are the homology and cohomology groups. Indeed, they're so useful that the study of their structure using **category** theory has become a branch of mathematics in its own right, and is known as **homological algebra**.

See also//
13 Categories, p.30
35 Abstract Algebra, p.74
38 Groups, p.80
49 Homological Algebra, p.102
60 Manifolds, p.124
70 Topology, p.144
76 Dimensions, p.156

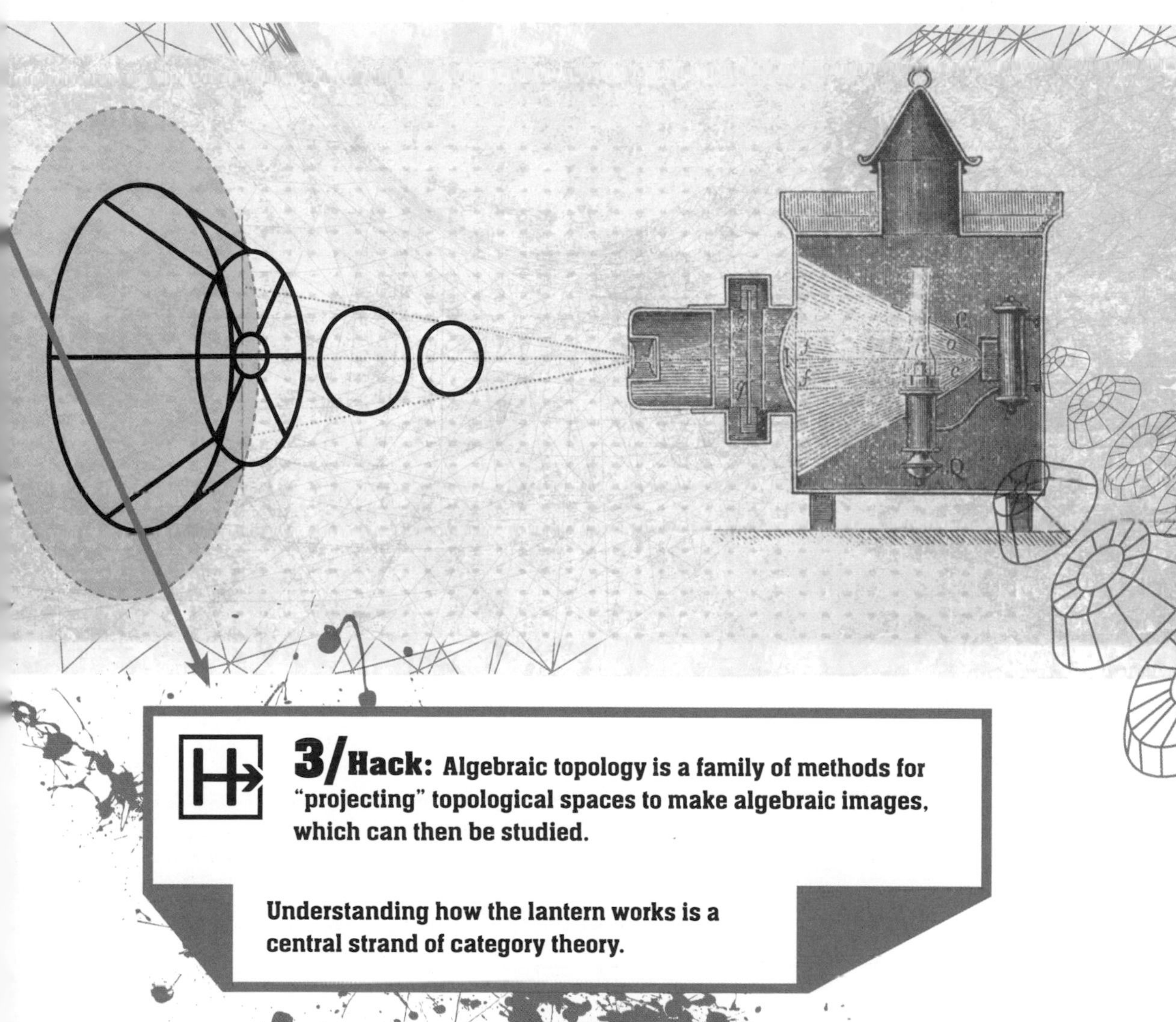

3/Hack: Algebraic topology is a family of methods for "projecting" topological spaces to make algebraic images, which can then be studied.

Understanding how the lantern works is a central strand of category theory.

No.85
Knot Theory Tying up loose ends

1/ Helicopter view: Interesting **topological** problems don't have to involve weird, unimaginable spaces dreamed up by pure mathematicians. Even simple everyday objects can pose puzzles when we look at them in the right way.

Humans have made knots for millennia. Any knot can be untied if its ends are loose and you have the patience, so mathematicians like to think of the ends of knots being joined together, making the piece of string into a circle.

In fact, topologically every knot is a circle. What makes a knot knotted is how it is embedded in three-**dimensional** space. In higher-dimensional spaces all knots fall apart, and in lower **dimensions** there is not enough room to tie them, so in a sense they are a specifically 3D phenomenon.

That said, there are analogs of knots in higher dimensions. In 4D you can tie a sphere in a knot, for example. *Knot theory* can also be extended to links and braids; the latter are important in particle physics.

Knots and links began to be classified in the 19th century, before the techniques of topology were available.

2/ Shortcut: Among other things, *knot* theorists ask how we can tell when two knots are essentially the same, meaning we can rearrange them to look identical without untying them. Much of this involves methods of **algebraic topology**.

Amazingly, every knot is the boundary of a surface (that is, a 2D **manifold**) that can exist in 3D space. These are mostly hard to imagine but a simple procedure exists for turning any knotted string into a paper model whose edge is the knot.

See also//
60 Manifolds, p.124
70 Topology, p.144
76 Dimensions, p.156
84 Algebraic Topology, p.172

3/ Hack: A knot is a circle embedded in 3D space; how it is embedded determines the knot.

Knottedness is a topological property of how the knot sits in space, not of the knot itself.

No.86
The Poincaré Conjecture
Zooming in on 3-manifolds

1/ Helicopter view: **Manifolds** form a special class of spaces that locally "look like" **Euclidean space** of some fixed **dimension**, but might be connected together in different ways on a large scale. A manifold is any space such that each point has a region around it that can be cut out and flattened so that it looks like a region of Euclidean space.

It's useful to "zoom in" on some special classes of manifolds. For example, a manifold is called *closed* if it has no boundary and doesn't shoot off to infinity. A square does have a boundary, which is **topologically** the same as a circle; the circle itself has no boundary, though, so it's a closed 1D manifold. Going up a dimension, we can imagine "inflating" a cube into an ordinary sphere, which is a closed 2D manifold.

Restricting further, we can concentrate on "simply connected" manifolds, which don't have holes like the one in a torus. The *Poincaré Conjecture* states that every closed, simply-connected 3D manifold is a 3-sphere. Grigori Perelman's proof was confirmed in 2006, at the same time as the closely related **Thurston's Geometrization Conjecture**.

Below: Topologically, a cube can be inflated into a sphere.

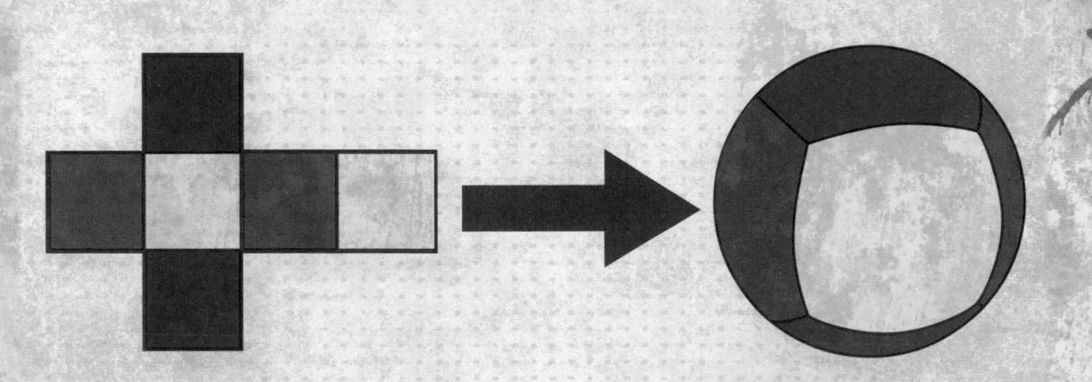

2/Shortcut: The things you can study in **topology** vary greatly depending on how many dimensions your space has. In five **dimensions** and above there is plenty of room for things to happen, causing many complexities that appear in lower dimensions to vanish.

The classification of 1D closed **manifolds** is easy because there is only one, the circle. In 2D the picture was complete by the 1920s. Perelman's results get us most of the way to classifying closed 3-manifolds. In higher dimensions, a logical blockage similar to **Gödel's Incompleteness Theorems** means no classification is possible.

See also//
6 Gödel's Incompleteness Theorems, p.16
59 Euclidean Spaces, p.122
60 Manifolds, p.124
76 Dimensions, p.156
81 Thurston's Geometrization Theorem, p.166
83 The Tesseract, p.170
84 Algebraic Topology, p.172

Jules Henri Poincaré (1854–1912)

3/Hack: A 3-sphere sounds exotic but in 4D space it is a very basic object; the Poincaré Conjecture tells us that all simply connected, closed 3-manifolds are 3-spheres.

4D objects such as the 3-sphere are not just theoretically interesting. In fact, Einsteinian spacetime is four-dimensional.

No.87 Varieties

Like manifolds, but weirder

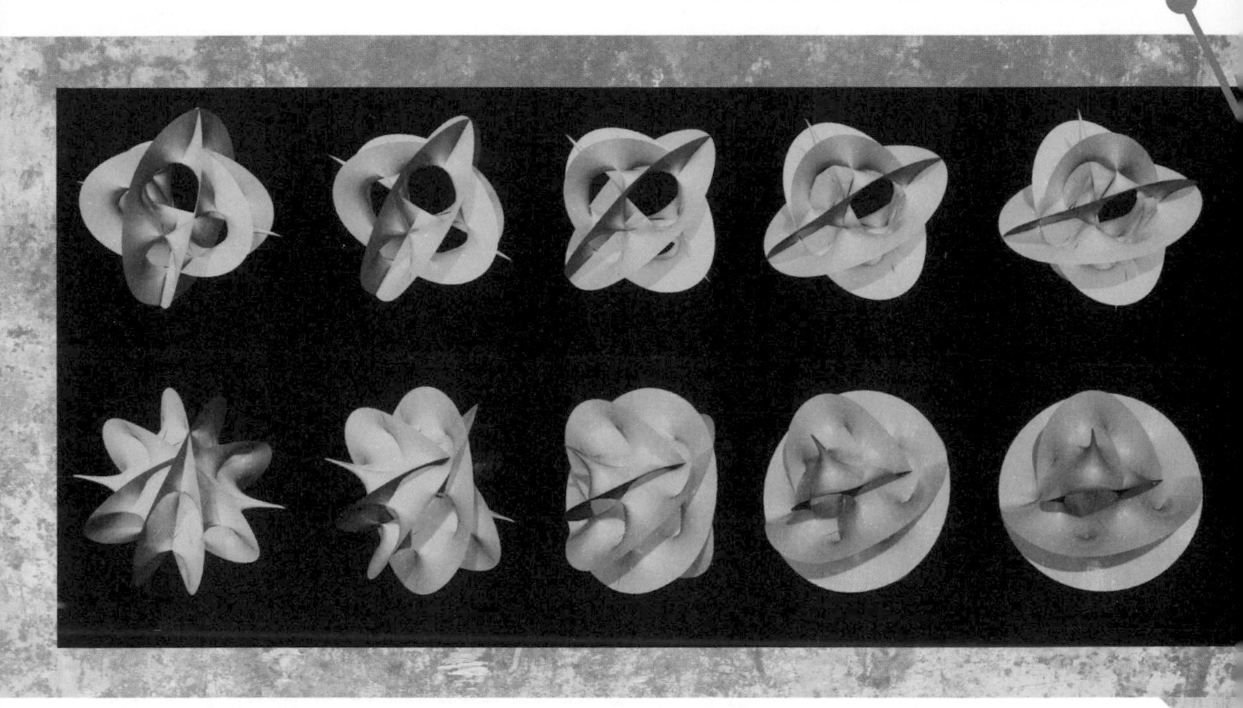

1/ Helicopter view: A graph is a handy way to visualize a **map** from a set of numbers to another set of numbers, but it can also be looked at as a geometric object in its own right.

Being a bit more rigorous, suppose we have a collection of **polynomials**, each of which has n variables. If we can find values for those variables that make all the polynomials in the collection evaluate to zero, we plot a point at those coordinates in **n-dimensional Euclidean space**.

The shape made of all those points is called a *variety*. Varieties can be lines, surfaces, or scatterings of points; they can also be far more complicated.

Polynomials in n variables belong to a **ring**, which is an abstract algebraic structure, so it is natural to try to use what we know about the algebra of rings to study the geometry of varieties. These are the roots of algebraic geometry, which is now a major field of mathematical research.

Varieties can exhibit local features that manifolds can't, such as self-intersections.

2/Shortcut: Modern algebraic geometry has been broadened out to include a wider class of objects, which are then amenable to a very abstract treatment that often yields powerful results.

The fundamental idea is that of a *scheme*, which is very similar to a **manifold**. Any small part of a scheme "looks like" an algebraic *variety*; the parts are "pasted together" using a structure called a sheaf. Sheaves and schemes are now the basic tools of the algebraic geometer's trade.

See also//

9 Maps, p.22
23 Polynomials, p.50
42 Rings and Fields, p.88
59 Euclidean Spaces, p.122
60 Manifolds, p.124
76 Dimensions, p.156

3/Hack: A variety is the set of points where a collection of polynomials all vanish to zero, while a scheme is an object that locally resembles a variety.

Varieties arise from graphs of polynomials. From this definition, a vast geometric universe arises.

No.88
Hilbert's Nullstellensatz
The algebraic geometer's dictionary

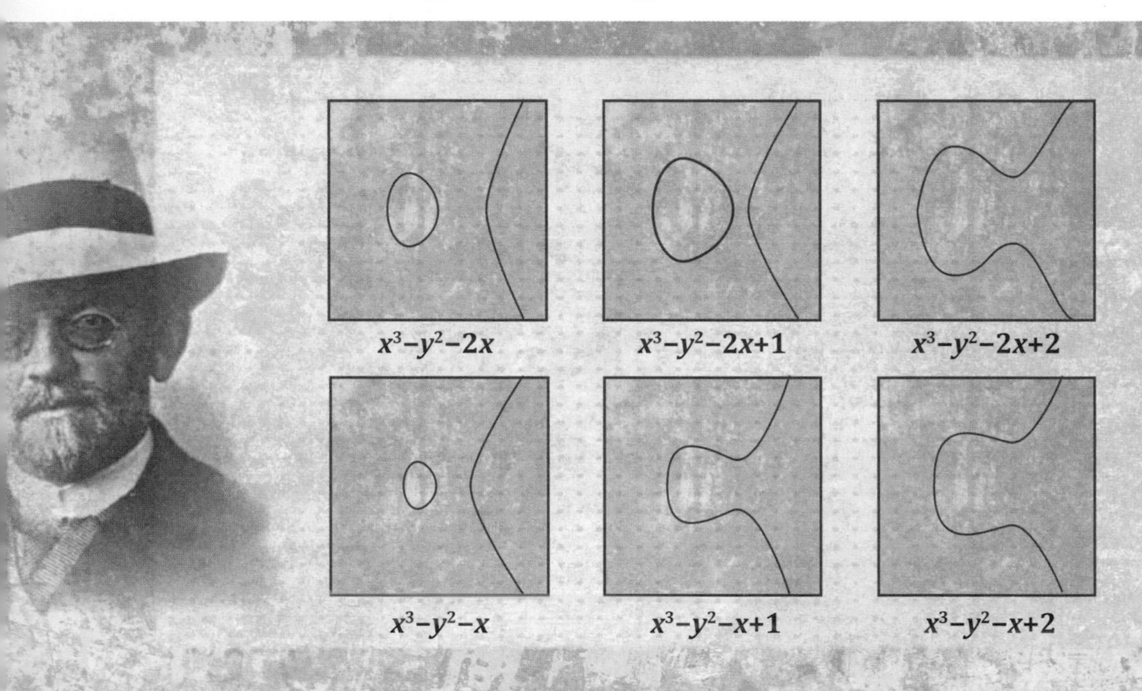

The variety associated with each of several polynomials, showing how the algebra changes the form.

1/ Helicopter view: Every **variety** has a set of **polynomials** in n variables that defines it. These in turn come from the **ring** of all the polynomials in n variables with coefficients in a particular **field**. So it is natural to think that studying how the defining polynomials sit inside their ring would be a good way to study the geometry of the **variety**.

In fact, it turns out there is a method to translate, in a very direct way, between *things we know about the algebra* and *things we know about the geometry*. This provides a kind of dictionary that allows us to pass between the two domains. So, when we get stuck on the geometry, we can translate the problem into algebra, then translate back when we are ready.

In the 1890s David Hilbert proved a result known as the *Nullstellensatz* ("the fact about zero points"), which establishes this correspondence in precise terms.

2/Shortcut: The *Nullstellensatz* says that there is a direct, one-for-one correspondence between **varieties** and things called *radical ideals*. A radical ideal is a subset of a **ring** that meets some simple requirements.

To every radical ideal corresponds one variety and vice versa. This is the heart of the Nullstellensatz, and it is what makes algebraic geometry possible.

As with algebraic topology, the point is that we know a lot about the properties of rings already, and so translating into that language is much more convenient than working with varieties directly.

See also//
23 Polynomials, p.50
42 Rings and Fields, p.88
87 Varieties, p.178

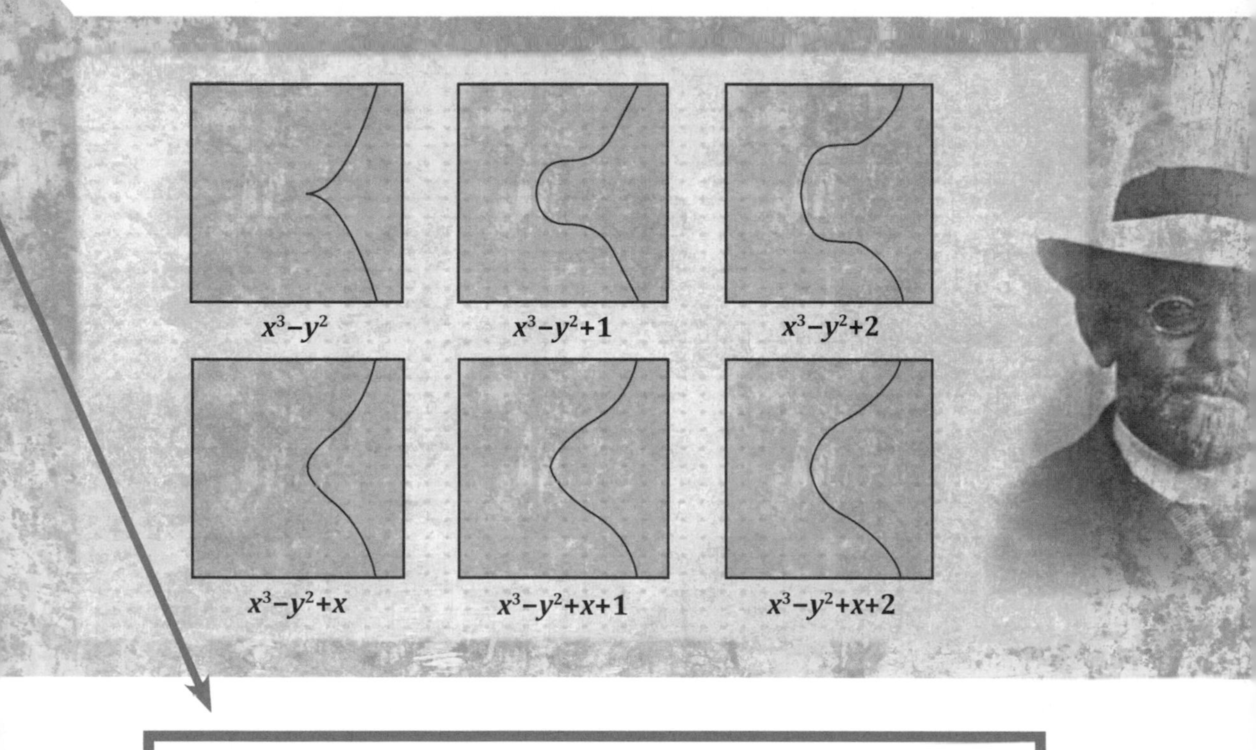

3/Hack: Varieties define ideals and ideals define varieties. This allows us to move seamlessly between the algebra of rings and the geometry of varieties.

The Nullstellensatz is the fundamental dictionary upon which algebraic geometry is built.

No.89
Iteration The power of repetition

1/Helicopter view: Consider slicing a loaf of bread. You start with a loaf and cut, say, a ½-inch slice off one end. Then you take the result of this operation—a loaf with ½ inch cut off—and repeat.

Carrying out the same procedure over and over, each time using the output of the last repetition as the input for the next, is an everyday example of *iteration*.

Mathematically speaking, iteration usually happens when you have a **map** whose domain and codomain are the same **set**. This allows us to apply the map repeatedly and see what happens.

It could be that an element of the set stays the same because the map sends it to itself. This is called a *fixed point*. Elements close to that point—assuming we have a **metric** that makes sense of that idea—may gravitate toward or away from it, in which case we call it a *sink* or a *source*, respectively.

Below: Iterative systems usually have a long-term behavior that can be categorized. *Right:* Many industrial processes iterate the same procedure many times to produce their final results.

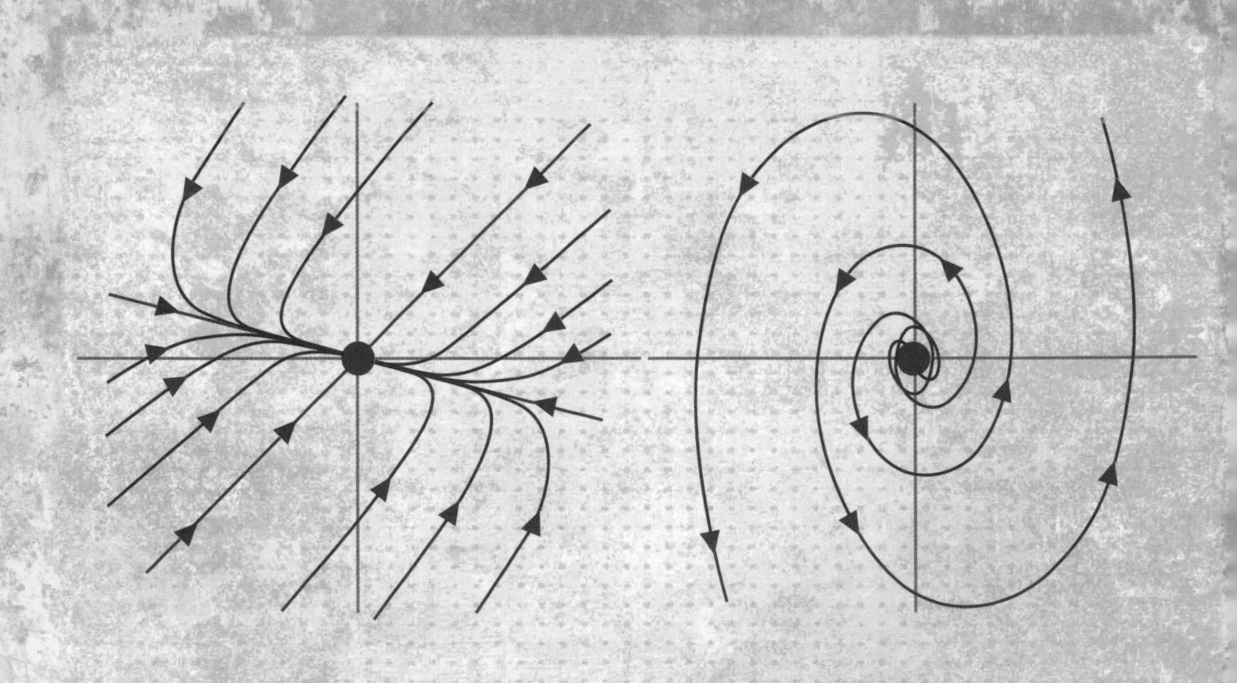

2/Shortcut: Suppose you have the set of all **rational numbers** and a **map** that sends each one to half of itself. If we *iterate* it and watch the number 8 the sequence starts: $8 \rightarrow 4 \rightarrow 2 \rightarrow 1 \rightarrow \frac{1}{2} \rightarrow \frac{1}{4}$. Here the number 0 is a fixed point, and it is a sink because other numbers nearby "gravitate" toward it.

If our map doubled numbers instead of halving them, 0 would still be a fixed point, but now it would be a source, since nearby numbers move away as you iterate.

See also//

7 Set Theory, p.18
9 Maps, p.22
15 The Collatz Conjecture, p.34
74 Metric Spaces, p.152
90 Brauer's Fixed Point Theorem, p.184
91 Chaos Theory, p.186

3/Hack: Iteration repeatedly applies a map whose domain and codomain are the same, using the output of each repetition as the input of the next.

Many natural, technological, and everyday processes are iterative.

No.90 Brauer's Fixed Point Theorem

Something always stays the same

L. E. J. Brouwer (1881–1966)

1/ Helicopter view: Stirring a cup of coffee is an **iterative** process. At every second, say, each point in the coffee cup is moved to a different point as it swirls around. The next second, this new state becomes the "input" for the next step of the stirring process. Think of each step as like a frames on a movie reel.

Brauer's Fixed Point Theorem says that, no matter how you stir your coffee, there is always a point that stays where it is: a fixed point.

This result is famous for two reasons, aside from how surprising it is. One is that it is not just about coffee cups, but applies to any suitable region of **Euclidean space**. Another is that continuous iterations and their fixed points are important in many areas of mathematics, especially the study of **differential equations**. Finally, it is fun to prove, and there are a great many different ways to do it.

Above: Any compact, disk-shaped region of space subject to a continuous transformation will have a fixed point.

2/Shortcut:

Brauer's theorem is true on any compact, convex region of **Euclidean space**. "Compact" means you could draw a box around the region (it does not go off to infinity) and it includes all its boundary points. A region of space is "convex" if, given any two points inside it, you can draw a straight line between them that stays inside the region.

Suppose you take a printed map that includes the place you are in, crumple it, and drop it on the floor. There will be a point on the map that is directly above the point it represents.

See also//

4 Limits, p.12
55 Differential Equations, p.114
59 Euclidean Spaces, p.122
89 Iteration, p.182

3/Hack: You cannot change everything all at once. Something, somewhere, always ends up staying the same (as long as the criteria for the theorem are met).

There is always a point on the Earth's surface where the wind is not blowing.

No.91 Chaos Theory

Complexity out of simplicity

1/ Helicopter view: There is no official definition of "chaos." In this context, it does *not* mean randomness or lack of order but rather something like "surprising complexity."

Often, complexity is unsurprising. Fluids, for example, are made of billions of independently moving particles, so when you drop a glass of milk on the floor we fully expect it do to something complicated.

Sometimes, though, complex behavior arises from simple systems. Think of a universe containing just three balls, and nothing else. They gravitationally attract one another; the physics is completely straightforward. But their behavior is unpredictable without exact information about their starting conditions. The tiniest nudge can produce wildly different results.

Both the falling milk and this "three-body problem" produce results that seem random but actually are not. The difference is that the latter involves only three simple objects and one equation, and even the equation does not have anything complex about it. Where has the complexity come from?

The Lorenz attractor, an iconic image from the early days of chaos theory, arose from the use of fluid dynamics to predict the weather.

2/ Shortcut: Chaotic systems generally have three things in common.

First, the behavior's complexity seems disproportionate to the simplicity of the rules governing it.

Second, small changes (for example, moving a planet 1mm from where it originally was) can produce completely different results (stable orbits instead of a head-on collision, say).

Third, the system is completely deterministic because if you had perfect information you could make perfect predictions. There is order in chaos; it is just hard to find.

Many chaotic systems are **iterative**. The results are sometimes **fractals**, the geometric objects that made the subject famous.

See also//

77 Fractional Dimensions, p.158

89 Iteration, p.182

97 Brownian Motion, p.198

3/ Hack: Many things that seem hopelessly hard to understand actually are not. With the right tools you can find surprisingly simple underlying structures.

Chaos is not randomness or disorder. It is a hidden order that allows complexity to arise from simplicity.

No.92 Factorials Exploding numbers

1/ Helicopter view: Suppose I have three songs I want to make into a playlist. How many ways can I do it?

Well, I can choose any of them as the first song, so that gives three possibilities. For the second I can choose any of the two remaining, so that offers two possibilities for each of the first three, making six in total. Finally, I have one song left, which I have to put into the final slot. This is how we calculate 3 *factorial*, written 3!. It is $3 \times 2 \times 1$, so $3! = 6$.

What if I add a fourth song? Now I have four choices for the first song, and then I am putting one of my three-song playlists on the end of that. This means I have $4 \times 3! = 4 \times 3 \times 2 \times 1 = 4!$ possibilities, which is 24. And so it goes on because you get the factorial of any **natural number** by multiplying together all the natural numbers less than it.

There are 355,687,428,096,000 ways to hang the pictures in this gallery.

2/Shortcut: *Factorials* get big fast. For example, 7! = 5,040, while 70! is far more than the number of atoms in the observable universe.

Many algorithms—roughly speaking, procedures that can be carried out by a machine—take a length of time proportional to the factorial of the number of things they have to deal with. A procedure might, for example, take 6 seconds for three pieces of data but 5,040 seconds for seven pieces of data; if you had 70 pieces of data, it would effectively take forever.

See also//
14 Natural Numbers, p.32
93 Combinatorics, p.193
100 P vs. NP, p.204

3/Hack: To find a natural number's factorial, multiply together all the natural numbers less than it.

Factorials "explode" extremely quickly, which can be a serious practical problem.

No.93 Combinatorics

The fine art of counting things

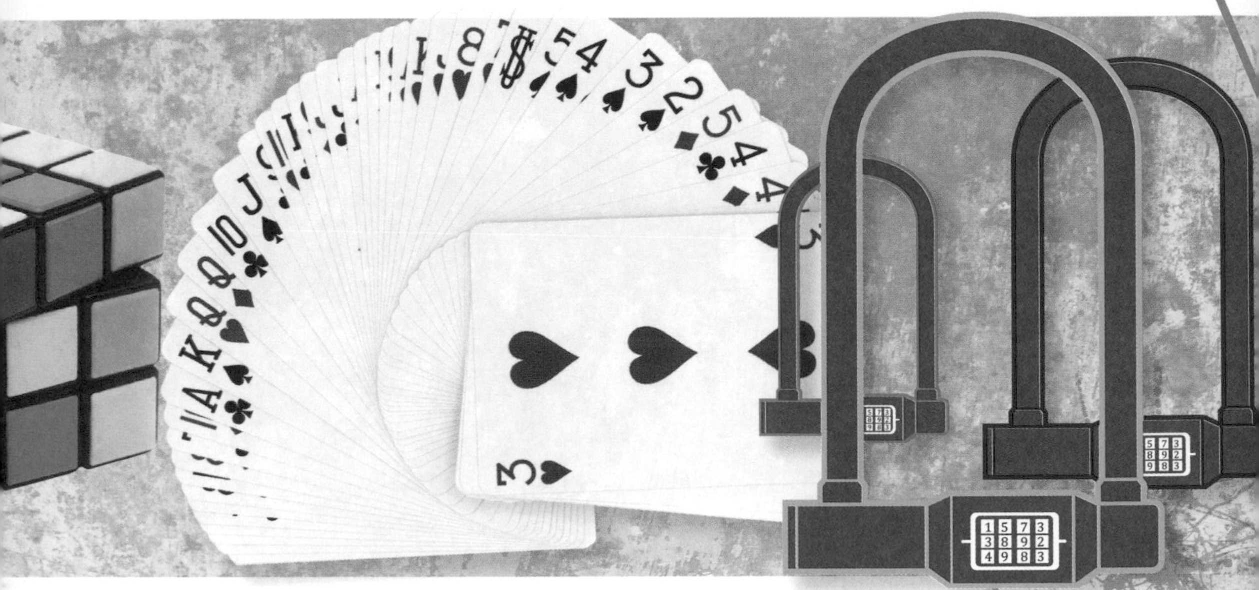

1/ Helicopter view: Mathematical questions need not be abstruse or technical. Some of the oldest take the form of a simple question, such as "How many are there?". The ancient mystical text *Sefer Yetzirah* includes solutions of several *combinatoric* problems using what we would today call **factorials**.

How many ways can a deck of cards be shuffled? How many configurations can a Rubik's Cube have? How many ways could we connect a dozen computers into a network? How many ways are there to write a number as a sum of integers?

Many problems like this result in huge answers, often because **factorials** are involved. Adding just one extra element to the problem can make the answer much, much bigger, a phenomenon is known as *combinatoric explosion*.

Mathematicians have developed a toolkit of cunning tricks for solving problems like these. The combinatoric techniques often have practical importance in themselves, but they are also used in unlikely-looking fields of mathematics such as **topology**.

Combinatorics arose from studying games but applies to many other areas, including cryptography.

2/Shortcut: Suppose I want to form a five-a-side soccer team from the 12 employees in my office. How many teams can I create? The problem is trickier than it sounds.

At first you might notice there are 12 choices for your first pick, leaving 11 for the second and so on, which gives $12 \times 11 \times 10 \times 9 \times 8$. We can express this as $12!/7!$.

But it turns out this answer is wrong; it counts each distinct team multiple times. The correct answer is $12!/(5! \times 7!)$, which is 792.

See also//

14 Natural Numbers, p.32

92 Factorials, p.188

94 Graphs, p.192

3/Hack: Combinatorics is the study of counting things, especially their possible combinations or permutations.

Counting can be hard, and combining even a small number of things can often be done in a huge number of ways.

No.94
Graphs Joining things together

1/ Helicopter view: In 1736 Leonhard Euler visited Königsberg. Its residents enjoyed taking circular walks that crossed as many as possible of its seven bridges. They preferred not to retrace their steps or cross a bridge more than once. The residents asked the great Euler, "Can there be a circular walk that crosses all seven bridges exactly once each?

Many problems are not really about the things they seem to be about, but about how those things are connected together. This is the general preserve of **topology**. We can work with a mathematical structure called a *graph*, which models the relations between objects. It is made of points called *nodes* joined together by lines called *edges*. Incidentally, these graphs happen to share their name with the curves we draw to visualize maps, but the two are unrelated.

Euler simplified a map of Königsberg to four points, one for each landmass, joined by seven edges representing the bridges. By doing this he was able to confirm what its residents had long suspected. No such walk is possible.

Below: Old Königsberg and Euler's graph that helped solve the puzzle.
Right: Eight different embeddings of the same graph; the differences don't matter, only the connections.

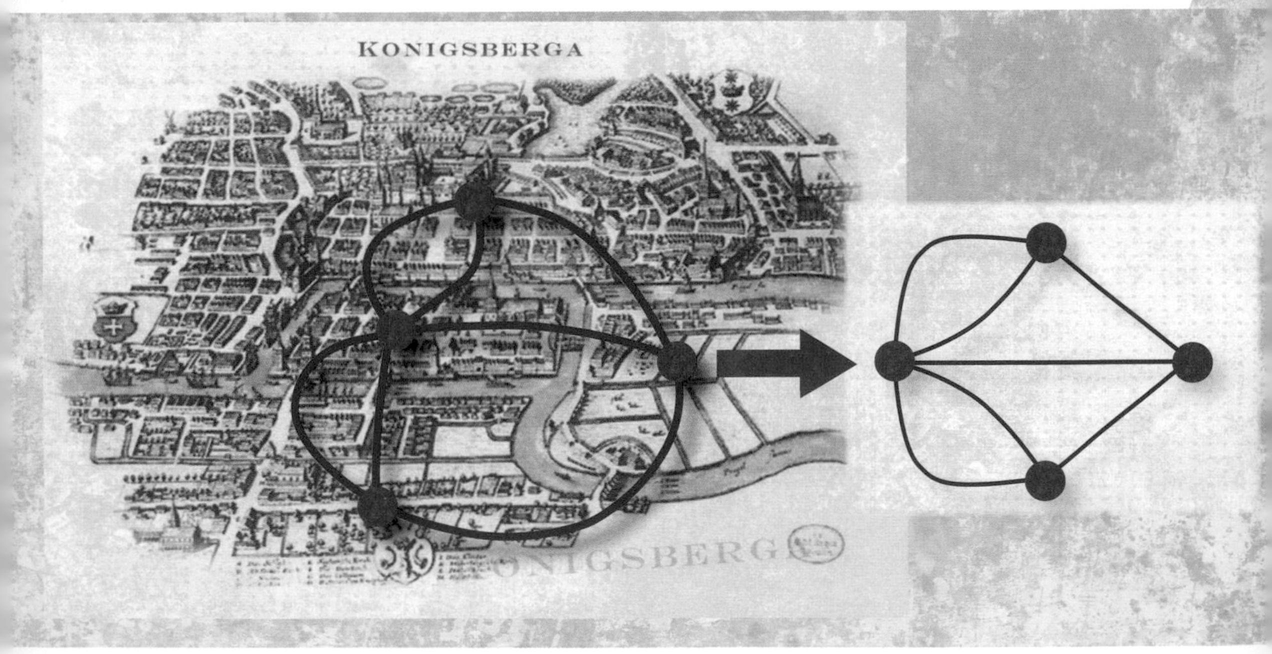

2/Shortcut: A *graph* is a very abstract object, and that is the point. Its nodes and edges can represent anything at all. Graph theory just tells us about the way they are connected together. Graphs are crucial in computing, algorithm design, logistics, decision-making, and more, as well as many fields of pure mathematics.

The edges of a graph can have numbers attached to them, making a *weighted graph*. They can also have an arrow added, making a *directed graph*. These allow them to be used in many other applications.

See also//
70 Topology, p.144
71 Triangulation, p.146
93 Combinatorics, p.190
100 P Vs. NP, p.204

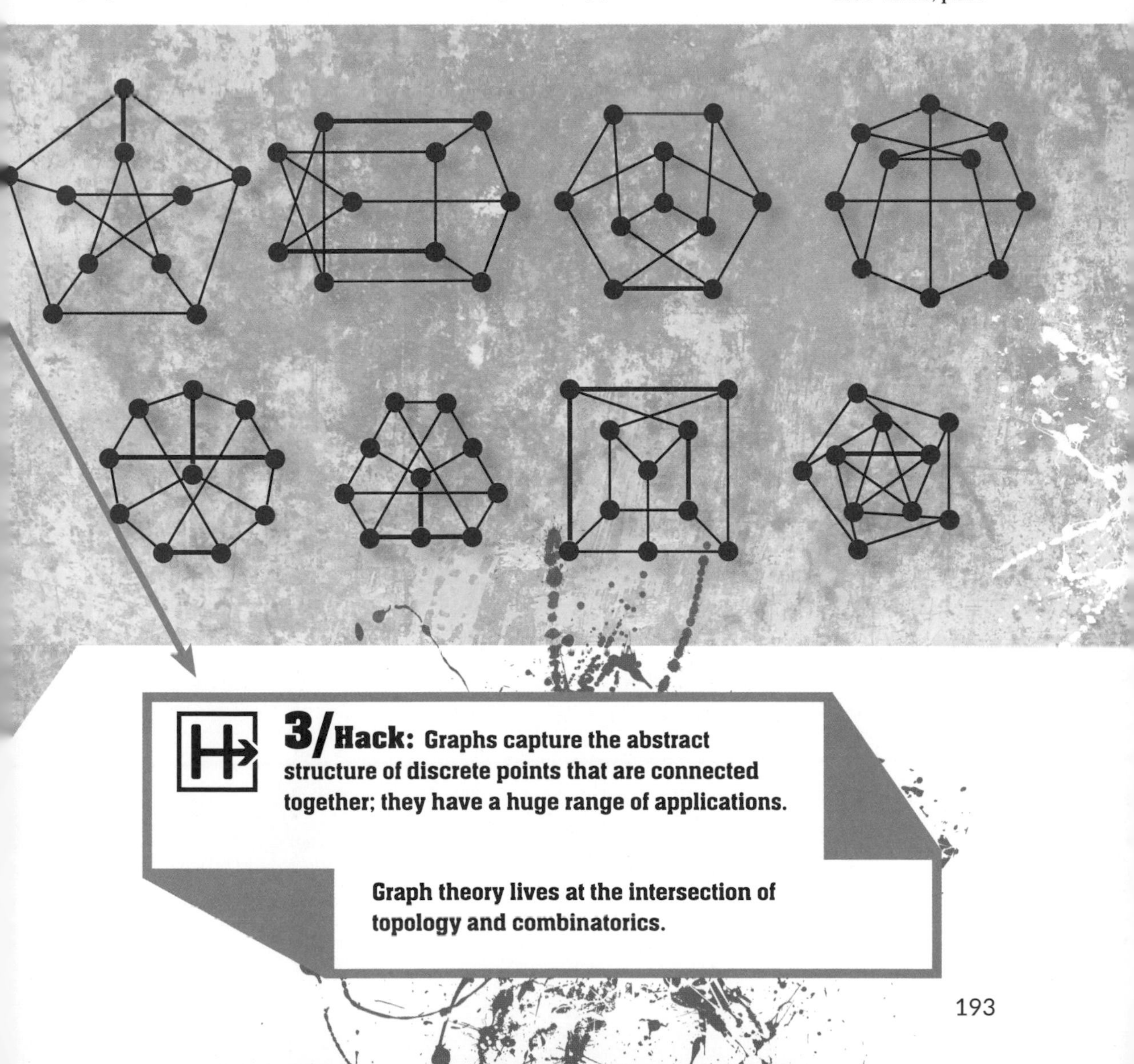

3/Hack: Graphs capture the abstract structure of discrete points that are connected together; they have a huge range of applications.

Graph theory lives at the intersection of topology and combinatorics.

No.95 Probability

Reasoning about uncertainty

1/ Helicopter view: Most mathematical theories start from **axioms** that we must accept as true. If I'm studying, say, **group** theory, I can assume that every group contains an identity element. If it didn't then it would not be a group, by definition.

In real life we rarely have truth-by-definition like this. More often we are uncertain. For example, when I roll two dice I know the total numbers they come up with will be between 2 and 12, but I have no way of knowing which (assuming I'm not cheating).

Does this mean I can't know *anything* about what number I might expect to get? It doesn't, and this is where *probability* comes in. Like **logic**, it offers a method of reasoning about the information we have. Unlike logic, it helps us deal with a range of possible outcomes without knowing which will happen in advance.

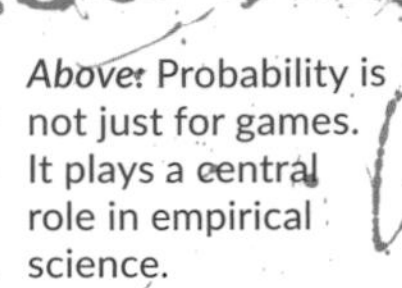

Above: Probability is not just for games. It plays a central role in empirical science.

2/ Shortcut: *Probability* was put on a firm footing by Andrey Kolmogorov in 1933, isolating the mathematics from philosophical questions about chance.

The idea is to form a set representing the possible outcomes, for example, all the numbers you can get from rolling two dice. You then assign a number to each subset representing the probability of what's in it actually happening.

Komogorov's **axioms** provide a simple algebra for combining these outcomes, so that you can ask complex questions like "If I roll two dice and neither comes up 6, what is the probability the total is more than 9?"

See also//
1 Axiom, Theorem, Proof, p.6
5 Logic, p.14
96 Statistics, p.196
97 Brownian Motion, p.198
98 Game Theory, p.200

Andrey Kolmogorov (1903–87) devised the standard axiomatic treatment of probability.

3/ Hack: Probability applies to situations where we have imperfect knowledge.

As with logic, probability is a formalized way of thinking.

No.96
Statistics A language for data

1/ Helicopter view: Suppose you want to learn about the heights of children in a school. You spend a happy day or two measuring the children, and carefully make a note of the height of each one in a long list. What have you learned? Not much. You've collected a lot of data but you need to interpret it if you want to learn anything.

Statistics is the name given to a toolkit of methods for making sense of data. Broadly speaking, these can be used in two ways: *descriptive* and *inferential*.

Descriptive statistics uses various averages and measures of how spread out the data is. We might ask for the average height of a child, or the greatest and least heights. There are more sophisticated examples too.

Inferential statistics usually looks at a sample, for example, the children at the school, and tries to infer information about a larger population, for example, all the children in the country.

Below: The Poisson distribution models events that happen at random intervals with fixed frequency, such as the arrival of buses at a stop.
Right: Statistics was first developed as a way for states to better understand their populations.

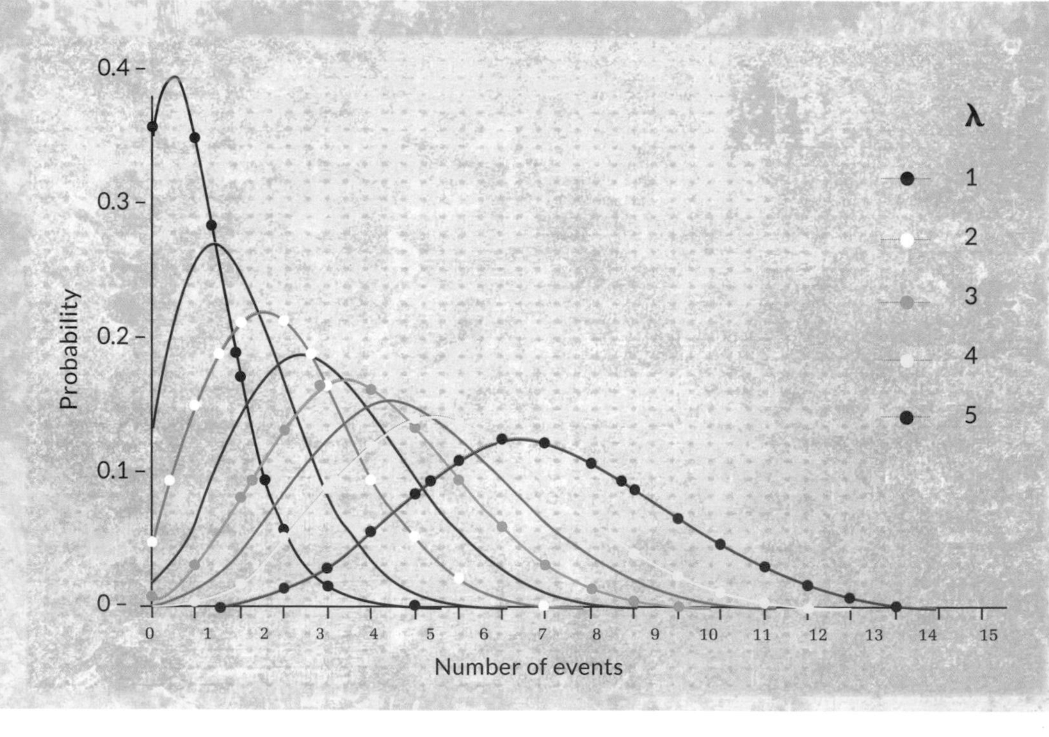

2/Shortcut: Although it uses mathematical methods, much statistical work falls outside the realm of mathematics itself, especially the inferential kind.

The present controversy over the statistic called a "*p*-value" is a case in point. Many scientific communities consider a *p*-value of less than 0.05 to constitute "statistical significance," making the difference between a meaningful finding and random noise. But there is nothing magical about 0.05 as a number. It is a threshold chosen because it seems reasonable.

See also//
95 Probability, p.194

3/Hack: Descriptive statistics includes average, range, and extreme values while inferential statistics seeks to draw reasonable wider conclusions that the data itself does not explicitly contain.

Statistics is a set of numerical tools for talking about data in a precise way.

No.97
Brownian Motion
Random walks with tiny steps

1/ Helicopter view: In 1827 botanist Robert Brown noticed that a pollen grain he was looking at under a microscope seemed to be moving around. Not only that, it kept changing direction, as if it was being jostled in a crowd. In 1905 Einstein give an explanation. The movement was caused by the pollen grain colliding with individual molecules of the water in which it was suspended.

There are a huge number of water molecules and each can only give a tiny nudge. What's more, these nudges are essentially random due to the intractable complexity of their behavior. So a good model for this "Brownian motion" would be movement that proceeds in imperceptibly tiny, unimaginably fast steps, each one in a random direction.

Anything that moves unpredictably in very small, frequent steps looks a bit like Brownian motion. Since the 1970s, for example, financial asset prices have often been modeled in just this way.

Random walks can be used to model share prices, atoms, and even human behavior.

2/Shortcut: Start with the idea of moving, say, 1m in a random direction each second. This is a "random walk," and is quite easy to study using the tools of **probability**. To get closer to *Brownian motion*, reduce both the size of the steps and the time we wait between them.

This leads us to the **limit** as the distance and time both become infinitesimally small. Surprisingly, it is possible to make mathematical sense of this, and the result is the Wiener process, a mathematical model for Brownian motion. Under certain assumptions, the Wiener process can be thought of as a **fractal** with dimension 1.5.

See also//
4 Limits, p.12
77 Fractional Dimensions, p.158
91 Chaos Theory, p.186
95 Probability, p.194

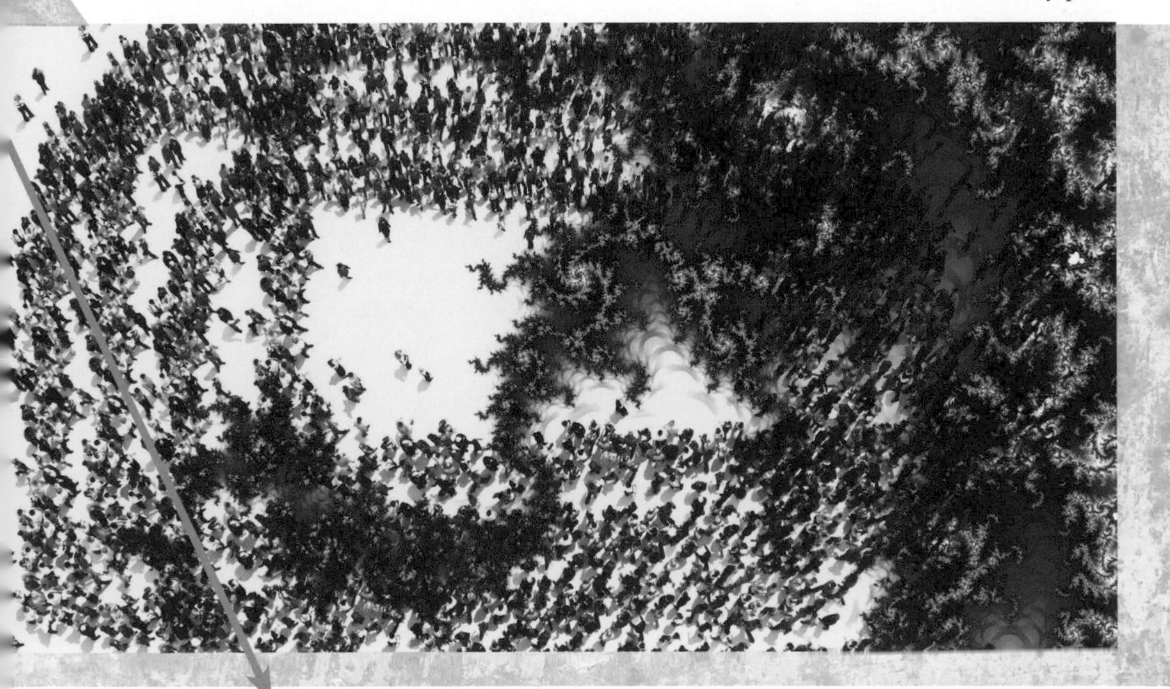

3/Hack: Brownian motion is a natural process; the Wiener process is a mathematical object that can be used to model it and other similar phenomena.

Brownian motion means moving in random, tiny, frequent steps.

No.98
Game Theory
Learning serious lessons from play

1/ Helicopter view: Alan and Betty have been arrested, but the police do not have any compelling evidence against them. If both keep quiet, the best the police can do is have them imprisoned for one year for not cooperating. The police want them to serve a longer sentence, so they offer them a deal to persuade one of them to give a statement incriminating the other.

If one sells out the other, the confessor goes free and the other party gets ten years in prison. If each betrays the other, though, each gets five years in prison. They are not allowed to communicate and each must decide alone. What should they do?

This is the Prisoner's Dilemma, a classic problem in *game theory*. It is about rational people seeking to maximize some quantity (here, time outside prison) with limited information.

Game theory is a framework for thinking about decision-making in situations where we do not know how others will act. It always deals in simplified situations, so real-world applications can be problematic.

Above: Game theory often focuses on games in which all players have the same information, such as chess.

2/Shortcut: The solution to the Prisoner's Dilemma is surprising.

If you are one of the prisoners, who cannot guarantee what the other will do, betraying is always the better choice. If you stay silent, and I betray you, I am saved from a year in prison for not cooperating. If you betray me, and I betray you, I am saved five years in prison. Either way, it is better for me to talk.

So each "should" betray the other, even though the best outcome arises from both remaining silent.

See also//
5 Logic, p.14
95 Probability, p.194

Prisoner B / Prisoner A	Prisoner B stays silent:	Prisoner B betrays:
Prisoner A stays silent:	Each serves 1 year	Prisoner A—10 years Prisoner B—goes free
Prisoner A betrays:	Prisoner A—goes free Prisoner B—10 years	Each serves 5 years

3/Hack: Game theory studies rational decision-making involving multiple choices, and often multiple people acting.

Game theory works very well in actual games, where its assumptions are often true.

No.99 Computability

What can a machine do?

1/ Helicopter view: The development of the general-purpose computer is one of the most significant technological events of the last century. Alan Turing was among the most important figures in its development, both theoretically and in practice.

Right: Turing's Bombe was used to break the enigma code during WWII.

To think about computers and their capabilities he invented a device now known as a Turing machine. This was never intended to be a practical computer, but it is easy to think about in a rigorous way.

Usually, a function (say) is called "computable" if, in theory, a Turing machine could be programmed to evaluate it. The Church–Turing Thesis claims that every function that can be carried out by an algorithm is computable, but this has not been proved (and it is not clear what it would mean to prove it).

Computability theory is still an area of intense research, and will probably remain so as long as we pursue the project of automating tasks using computers.

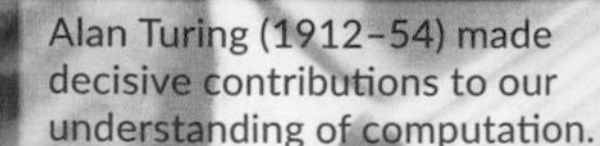

Alan Turing (1912–54) made decisive contributions to our understanding of computation.

2/ Shortcut: A classical Turing machine consists of a tape (as long as needed) and a "head" that can both read a symbol into the machine's memory and write one onto a blank section of tape. The machine has a limited number of memory slots for stored symbols (called "registers") and a second memory where it holds instructions telling it what to do (the "program").

This physical setup is just an example, however. Every laptop, smartphone, set-top box, and so on has exactly the same powers as a Turing machine.

See also//

6 Gödel's Incompleteness Theorems, p.16

69 Impossible Constructions, p.142

100 P vs. NP, p.204

3/ Hack: A Turing machine is a simple device that serves as a formal model of a computer. Any real system capable of doing everything it can is called "Turing complete."

The boundaries of computability are the limits of what a computer can do.

No.100
P vs. NP
Some jobs cannot be rushed

1/ Helicopter view: Computational complexity theory studies how difficult it is to carry out algorithms (procedures that can be carried out by a machine). In particular, it looks at how much longer an algorithm will take if you increase the number of things it has to work on.

Suppose you want to find the oldest person in a group of ten people. All you have to do is ask each person their age, and at each step keep track of who is oldest so far. After ten steps, you know who is oldest. If you had a hundred people, it would take a hundred steps.

We say this problem can be solved in *linear time*. Many problems cannot be solved in linear time, but can still be done in a time that is some **polynomial** function of how many things they have to deal with. They are all members of a class called *P*.

For some problems, however, we only have algorithms that are much less efficient than this. They belong to a class called *NP*.

A salesman plans to visit several cities. Which route is most efficient? This is believed to be an NP problem.

2/Shortcut: With a lot of data, an *NP* problem is a real nightmare. Instead of running in a few seconds it might take hours or days. Often these problems are of practical importance, prompting a search for better algorithms.

But can we always do better? Is a problem NP only because we haven't found the "right" way to solve it yet? That is, is NP really *P*? Or are some problems intrinsically too hard to solve in **polynomial** time? At the time of writing, we don't know.

See also//
23 Polynomials, p.50
92 Factorials, p.188
99 Computability, p.202

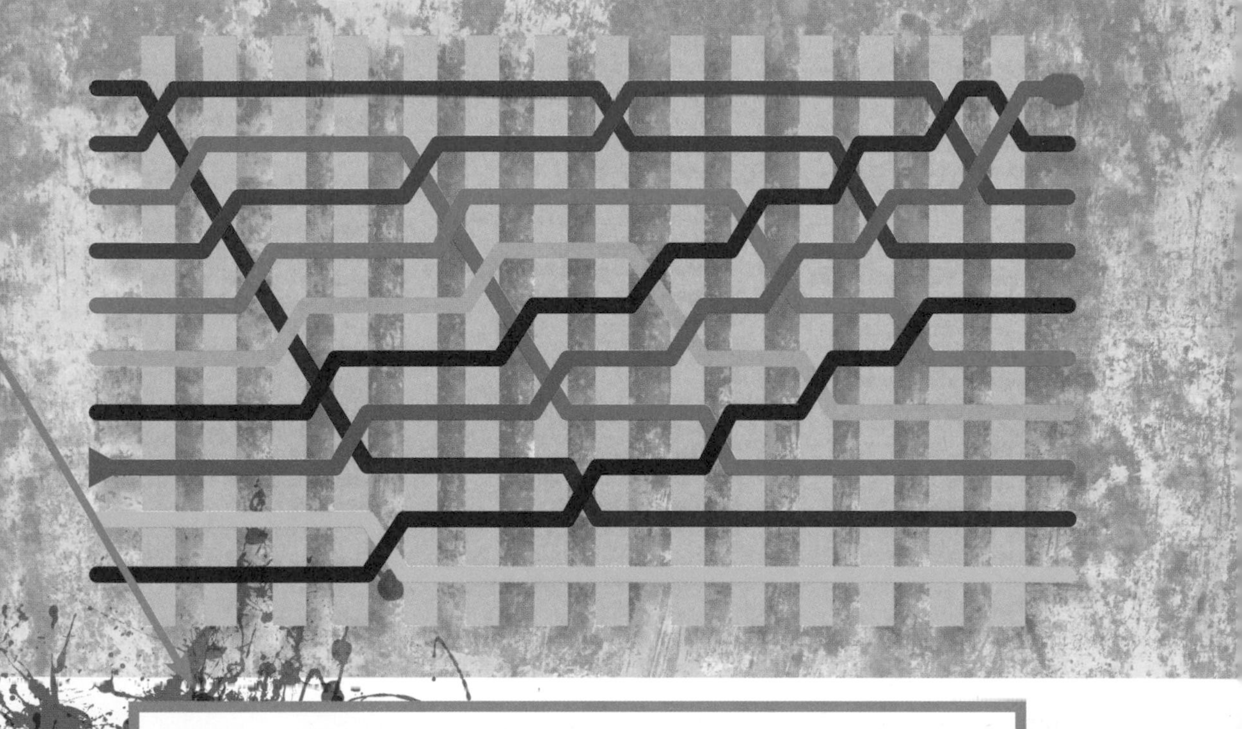

3/Hack: Polynomial-time (P) algorithms are practical even for quite large amounts of data; nonpolynomial (NP) ones are not.

Are problems NP only because we have not found a clever way to solve them, or is that impossible?

Index

Acknowledgments

Alamy Stock Photo 507 collection 168l & c; 914 collection 86l; ART Collection 112; Christopher Jones 66; Chronicle 6al, 108, 177; dpa 12; Everett Collection Historical 197; Gibson 168r; Granger Historical Picture Archive 18l, 53cl; Ian Paterson 37; Interfoto 14br; Laguna Design/Science Photo Library 178-9; Louis Berk 203; Paul Fearn 6cr; Photo Researchers/ Science History Images 65c, 150r, 202r; Pictorial Press 16l; Prisma Archivo 6cl; Sputnik 195; The Granger Collection 94, 98l; **Dreamstime.com** Almagami 16r; Andreykuzmin 23c; Angelo Cordeschi 136l; CarolAnneFreeling 74; Eldadcarin 136r; Ermolenko 20l; Everett Collection Inc 33; Idorum 20r; Jinfeng Zhang 188; Jonathan Lingel 15al; Mykola Lytvynenko 165l; Nicku 23 inset; Nicolastico 184r; Numgallery 8b; Peeterson 23a; **Getty Images** Hank Morgan/The LIFE Images Collection 158l; John Prieto/The Denver 9; Richard Hartog/Los Angeles Times 41; **Library of Congress** Prints and Photographs Division 72, 53cr; **https://math.stackexchange.com** Mariano Suárez-Álvarez (CC-BY-SA 3.0) 30; Image courtesy of **Jens U Nöckel** 133; **REX Shutterstock** AP Photo/Charles Rex Arbogast 95; **Rijksmuseum, Amsterdam** 169; **Science Photo Library** Bodleian Museum/Oxford University Images 7r; Royal Astronomical Society 7l&c; **Shutterstock** 120; Pataporn Kuanui 119; aaltair 52; Aida Pacheva 172c, 173r; Alena Ozerova 144r; Alex_Po 141r; andante 171l; Andreas Rauh 10bc; Anna Poguliaeva 82; anucha sirivisansuwan 154; Benjamin Haas 64l; bernashafo 185; bibiphoto 29; Birdiegirl 46r; bluebay 198; Brian J Abela 116l; brovkin 145 background; Business stock 138c; cgterminal 144l, 174r; Chomphuphucar 121; ClickHere 156-157; concept w 13a; Dario Sabljak 142; Daxiao Productions 54c; dikobraziy 124r; Dima Moroz 145c; donatas1205 14cr; elfinadesign 17c, 159c; Elizabeth Scofidio 166 background; Elnur 37c; ESB Professional 200; Everett Collection 183; Evlakhov Valeriy 96; fizkes 80; freesoulproduction 140; gabydesign 31; Georgios Kollidas 148c; Geza Farkas 60l; Grand Warszawski 150l; Guten Tag Vector 175; Hein Nouwens 48c; Ian Dikhtiar 190c; In Green 118; Jennifer Gottschalk 71; Jurgen Ziewe 202l; Juriaan Wossink 53r; Khanchit Kamboobpha 138 background; koi88 199; kovalto1 163c, 171r; LongQuattro 10br, 18r, 25, 166a; majivecka 28r; Marza 151; Merfin 134; Mic hael 130; Michal Sanca 28c; milart 135; milyana 104l; Mopic 159l; Morphart Creation 173c; nicemonkey 6ar, 17b, 21b, 23b; nobeastsofierce 38l; NShu 45r; Pabkov164; Petrovic Igor 100r; Pupes 49; Repina Valeriya 146; RFvectors 160; rk graphic 166c; Robert Varga 131; RoboLab 204; Rozilynn Mitchell 15bl; RRong 19r; SP-Photo 137; Stefan Mlynarcik 152; stocker1970 162; str33tcat 58c; Sylverarts Vectors 149l&c; syzius 20c; Tamara Lucic Dinic 11; theerapol sri-in 191; Tony Baggett 114c; Umberto Shtanzman 124l; Vastram 190l; VectorWeb 186; vertical 86r; Vetreno 33b; Vidya Thotangare 147r; vincent noel 4; Wetzkaz Graphics 194; Yorik 116r; Yuliyan Velchev 13c; Yurii Andreichyn 148l, 174l; **Wellcome Collection** 106l, 8a, 53l; **Wikipedia Commons** Adam Cunningham and John Ringland (CC-BY-SA 3.0) 43; Cronholm144 (CC-BY-SA 3.0) 54r, 58r; Empetrisor (CC-BY-SA 4.0) 98c; Geek3 (CC-BY 3.0) 59r; photo by Konrad Jacobs (CC-BY-SA 2.0) 34l, 100l; Lars H Rohwedder, Sarregouset 161; Petrus3743 (CC-BY-SA 4.0) 143; Wolfkeeper (CC-BY-SA 3.0) 70r.